Yale Studies in the History of Science and Medicine, 11

Conical Sundial from Delos, Greece

Greek and Roman Sundials

Sharon L. Gibbs

Yale University Press, New Haven and London

1976

Library of Congress catalog card number: 75-18173
International standard book number: 0-300-01802-9

Designed by John O. C. McCrillis.
Printed in the United States of America by The Murray Printing Co., Westford, Massachusetts, 01886.

Published in Great Britain, Europe, and Africa by Yale University Press, Ltd., London. Distributed in Latin America by Kaiman & Polon, Inc., New York City; in Australasia by Book & Film Services, Artarmon, N.S.W., Australia; in Japan by John Weatherhill, Inc., Tokyo.

Permission to reprint photographs has been granted by the following: plates 11 and 39 by the Département des Antiquités grecques et romaines, Musée du Louvre, Paris; plates 15 and 19 by the Deutsches Archaeologisches Institut, Rome; plate 29 by the Corinth Excavations of the American School of Classical Studies, Athens; plate 38 by the Department of Antiquities, Nicosia; plate 42 by the Kunsthistorisches Museum, Vienna; plates 48 and 59 by the Trustees of the British Museum, London; plate 49 by the Département des Antiquités orientales, Musée du Louvre, Paris; plate 53 by the Museo Archeologico, Aquileia.

Contents

Acknowledgments

The preparation of this catalogue of Greek and Roman sundials would not have been possible without the generous financial assistance of the Archaeological Institute of America's Olivia James Traveling Fellowship. Institute support was helpful in other ways in that it encouraged the cooperation of archaeologists and opened many museum doors. As a result of my association with the AIA, I was able to benefit from the experience and knowledge of professors and librarians at the American Academy in Rome, the American School of Classical Studies in Athens, and the Dutch Archaeological Institute in Istanbul.

Any list of individual acknowledgements is inevitably too short. Mine must certainly include Professors Asger Aaboe, Bernard Goldstein, Otto Neugebauer, and Derek de Solla Price, all of whom provided notes, suggestions, and encouragement at appropriate moments. Mr. L. Janin kindly provided me with information on all the Gallo-Roman dials known to him. Mr. and Mrs. R. S. Webster also shared information about dials they had seen. While neither is to be held responsible for errors in the final edition, both Professor Gerald Toomer and Dr. Cynthia Thompson gave much needed assistance in interpreting Greek and Latin inscriptions and sources. Mrs. Patricia Ryan, with the occasional assistance of Mrs. Helen Payne, prepared a virtually flawless press-ready copy of my manuscript.

I am grateful for the assistance of curators in about seventy museums in Europe and Central Asia, who allowed me to examine and photograph preserved dials. Occasionally their hospitality extended to the provision of nourishment and first aid. Seven institutions very kindly gave permission to reproduce photographs from their files: the Museo Archeologico, Aquileia, Italy (plate 53); the Corinth Excavations (plate 29); the British Museum, London (plates 48 and 59); the Department of Antiquities, Nicosia, Cyprus (plate 38); the Musée du Louvre, Paris (plates 11, 39, and 49); the Deutsches Archaeologisches Institut, Rome (plates 15 and 19); and the Kunsthistorisches Museum, Vienna (plate 42).

The patience and encouragement of the editors of Yale University Press has seemed endless. Their assistance has greatly reduced the number of errors possible in typewriter composition. Any errors which remain are, of course, my own responsibility.

Greek and Roman Sundials

1 The Evidence

> Ἀντίον ἠελίου στήσας ῥῖνα καὶ στόμα χάσκων,
> δείξεις τὰς ὥρας πᾶσι παρερχομένοις.
>
> If you put your nose pointing to the sun
> and open your mouth wide,
> You will show all passersby the time of day.
>
> *Anthologia Graeca 11.418*

Scientific archeological excavations of Greek and Roman sites have frequently led to the discovery of at least one --and often more than one--stone object which, having been labeled "sundial," is stored in a study collection among the minor architectural fragments. These sundials may justifiably be classified among the smallest pieces of Graeco-Roman architecture, but they do form the largest part of the meagre physical evidence for the scientific activity of the period. What little other physical evidence for scientific activity in Greek and Roman times is provided by a small group of metal objects. The group includes a bronze geared calendrical device found off the coast of Antikythera, Greece, and dating from about 80 B.C. To this sophisticated mechanism may be added a few portable sundials and some surveying tools and drawing instruments.

Heretofore the usefulness of the evidence provided by Greek and Roman dials has been limited by incomplete and inconsistent presentations. The dials are now preserved at archeological sites around the Mediterranean as well as in numerous European museums. Many have not been reported in the literature. The descriptions which have appeared in journal articles and excavation reports often give very little information. One exception is the discussion of sundials found on Delos published in volume 18, *Le Mobilier Délien*, of *Exploration Archéologique de*

Délos, Paris, 1938. This article by W. Deonna contains useful, consistent descriptions, measurements, drawings, and related bibliography.

In the early part of this century, two German scholars, Hermann Diels and Joseph Drecker,[2] presented some of the then available published material in valuable comparative studies of ancient dials. It is clear from their contribution that an adequate analysis of ancient dialling requires complete information gathered in a comprehensive catalogue. A complete, consistent catalogue of 256 known Greek and Roman sundials, including unpublished examples, is here presented for the first time. It represents an effort to consider all the available archeological information on the work of the ancient dialler.

Nonportable dials from Graeco-Roman antiquity were carved from blocks of different kinds of calcareous stone (marble, limestone, tufa). The largest preserved example, excluding the famous Tower of Winds in Athens, is 977 mm (3.2 ft) in height; the smallest complete example is 65 mm (2.5 in) in height. Most examples are about 300 mm (1 ft) in height. Whatever their dimensions, ancient Greek and Roman sundials are easily distinguished from modern garden varieties by their shape and, more importantly, by their engraved network of hour lines and day curves. The evidence suggests that Greek and Roman diallers preferred conical and spherical dial surfaces. Not one disk-shaped horizontal model is known. On modern dials a number of hours of equal length throughout the year are usually marked on a horizontal surface by the shadow of the *edge* of a gnomon which lies parallel to the axis of the earth. Greek and Roman sundials always marked the twelve seasonal hours[3] of daylight between sunrise and sunset. But while the seasonal hours were of equal length during a given day, their length varied during the year, being shortest at winter solstice and longest at summer solstice. On a Greek or Roman dial, the shadow of a gnomon's *point*, not its edge, marked the seasonal hours. The base of the gnomon could be placed in any convenient position. A glance at the catalogue shows that only a few gnomons are preserved. These and traces on other examples indicate that Greek and Roman diallers fixed bronze or iron gnomons, shaped like elongated pyramids, in lead-filled holes on the dial face somewhere along the meridian line.

Even if the pointed end of a Greek or Roman dial's gnomon is not preserved, its original extent can usually be found quite easily by examining the network of hour lines and day curves engraved on the dial face. With few exceptions, the day curves trace the path of the shadow of the gnomon's point at solstices and equinoxes (three lines in all), so the accurately engraved sundial served as a crude calendar. Its calendar day curves are divided by eleven hour points and connected by eleven hour lines. The first hour begins at sunrise; the last hour ends at sunset. Hence, when it is not itself a horizontal plane, the effective shadow-receiving surface of a Graeco-Roman dial is below and bounded by a horizontal plane which passes through the gnomon point. The frontispiece illustrates a typical Roman dial from Delos (catalogue number 3024G) with conical south-facing shadow-receiving surface and preserved gnomon in position. The shadow of its gnomon point lies on the equinoctial line between the fourth and fifth hour points.

The earliest Greek dial included in this catalogue dates from the third century B.C.; the latest Roman dial (not Byzantine) dates from the fourth century A.D. Remaining catalogue entries refer to dials which were made in the period between these extremes. It is difficult to determine an exact date of origin for most examples. Few were found in their original location. Many were found being reused as construction material. Most lack datable inscriptions or ornamentation.

Literary Sources

Centuries of excavations at the site of Pompeii have uncovered all or part of some thirty stone sundials. The dials were found in courtyards of private homes, in public baths, near temples, and in public squares. Citizens of that city must have been very familiar with the image of a pointed gnomon casting shadows on a grinning grid. This kind of familiarity would have enabled them to appreciate the epigram attributed to Trajan which opens this chapter. All of the types of dial uncovered at Pompeii are mentioned in book 9, chapter 8, of Vitruvius's *De architectura* which probably was published about a century before the fatal eruption of Mt. Vesuvius.[4] Vitruvius's brief and sometimes obscure remarks are yet the most complete on the subject of dialling in an extant ancient source. We are in debt to him not only for a list of the types of sundials in use in his day,[5] but also for a description of a geometric method used in their construction.[6] Vitruvius clearly specified the kind of time a dial would show passersby (9.7.7): "uti dies aequinoctialis brumalisque idemque solstitialis in duodecim partes aequaliter sit divisus."

Literary and archeological evidence can easily be reconciled to identify the role and character of sundials in use in the first century A.D. The task of tracing the origins and early development of eventually standard types is a much more difficult one, complicated by ambiguous literary evidence and a disappointing lack of early archeological data.

Various popular Greek and Roman authors have referred to the introduction of certain instruments into Greece which have been traditionally identified as the first Greek sundials. The earliest (430 B.C.) and most quoted source is Herodotus (2.109.3) who states that the Greeks acquired the twelve parts of the day and the πόλος and γνώμων from Babylonia.[7] The statement is a curious digression in a book devoted to a description and history of Egypt. The extraordinary nature of the remark has been emphasized most recently by the work of O. Neugebauer and R. A. Parker, who have shown that the origins of the seasonal hours are more likely to be found in the ancient Egyptian (1800-1200 B.C.) practice of reckoning nighttime during the year according to the rising of decans.[8] The puzzling nature of Herodotus's remarks has led at least one classicist to doubt their attribution.[9] Mentioned in connection with the twelve parts of the day, the πόλος and γνώμων could very well be timekeepers.[10] Unfortunately we have no clear description of their appearance or use.

A much later Greek author gives more information about the γνώμων referred to by Herodotus. Diogenes Laertius (310-320), quoting Favorinus (second century A.D.), credits Anaximander (580 B.C.) with first inventing a γνώμων and then setting it up ἐπὶ τῶν σκιοθήρων at Sparta.[11] The two terms γνώμων and σκιοθήρης may be used synonymously here. Vitruvius (1.6) assures us that they refer to the same instrument. Σκιοθήρης means "shadow catcher." It calls to mind an instrument whose function depends on the position of the sun. Further comments in Laertius suggest that Anaximander's γνώμων was used to observe solstices and equinoxes. Hours are not mentioned, and it is probable that this device did not mark them. The instrument may have consisted of a pointed shadow-caster fixed perpendicularly to a straight meridian line, engraved on stone and divided according to the noon shadow lengths at various times of year. In 1938, D. W. S. Hunt found an inscribed stone on Chios, dating from the second century B.C., which satisfies this description.[12]

In addition to γνώμονες, Anaximander is credited by Laertius with the construction of ὡροσκοπεῖα.[13] From a second reference in *Lives of Eminent Philosophers* (6.104) we learn that the Greek biographer considered the ὡροσκοπεῖον a timekeeping device. He relates an anecdote about the Cynic Diogenes of Sinope (403-322 B.C.), who, when presented with a ὡροσκοπεῖον, recognized it as a practical instrument which enabled one to arrive on time for dinner. References to ὡροσκοπεῖα in other sources (Strabo 2.5.4 and Geminus 2.35) suggest that their method of keeping time involved cast shadows. For final confirmation of the meaning of the term we can refer to the corpus of preserved sundials. A conical dial (catalogue number 3055G) now in the Pergamon Museum in Berlin is clearly referred to in its dedicatory inscription as ὡροσκόπιον. The Cynic may indeed have been shown a sundial in the fourth century, and Anaximander may have introduced the instrument into Greece in the sixth century.

References to γνώμονες, σκιοθῆραι, ὡροσκοπεῖα, and other possible early sundials[14] place these instruments in the hands of scientists and philosophers and give little indication that they were used by the average citizen. In fact, we have reason to believe that, in pre-Hellenistic times, most Greeks were not interested in marking the seasonal hours. Two similar anecdotes, attributed to Eubulus (370 B.C.) and Menander (315/14 B.C.), and a line from Aristophanes' *Ecclesiazusae* (393 B.C.) indicate that meal times in the fourth century B.C. were specified by the length of the hungry person's shadow (σκιά, στοιχεῖον).[15] The practice is described and discussed in detail by Gustav Bilfinger in *Die Zeitmesser der Antiken Völker*--especially chapter 2. He considers the use of this method evidence for the unpopularity of seasonal hours in pre-Hellenistic times.[16] There are other indications that this early period lacked the kind of standardized timekeeping provided by sundials. In *Epidemics* 7, a part of the Corpus Hippocraticum probably written on Cos about 360 B.C., time is indicated by referring to activities in the agora rather than to shadows on a sundial.[17]

On the whole, the literary evidence seems to suggest that there were few, if any, sundials marking the seasons and seasonal hours in Greece before the third century B.C.

Archeological evidence confirms this hypothesis. We do know that the notion of seasonal hours has a long history, and we know that classical astronomers used instruments which marked seasonal changes in the position of the sun. These practices may well have combined to inspire the design of the sundial, perhaps as early as the sixth century, but preserved examples thought to date even from Hellenistic times are few. However, Vitruvius indicates that many types of dials were invented in the postclassical period.

The Scientific Context

Vitruvius's list of sundial designs includes the names of their inventors. The more familiar names are those of Hellenistic astronomers and mathematicians: Apollonius, Aristarchus, Berosus, Dionysodorus, Eudoxus, Theodosius. This impressive roster suggests that the sundial developed as a scientific instrument in the fourth and third centuries B.C. There are several reasons why the successful designer (not to be confused with the maker) of dials would have to possess the talents of both a mathematician and an astronomer. As astronomer he would consider the observable path of the sun. As mathematician he would contrive to project that path onto a shadow-receiving surface.

In book 1, chapter 6, of the *Almagest*, Ptolemy considers some facts about the celestial sphere, established by Hellenistic astronomers, which were certainly of particular interest to astronomer-diallers in Graeco-Roman antiquity. He points out that the size of the earth is negligible compared to the diameter of the apparent surrounding sphere containing the sun's path.[18] Therefore, (1) the point of a gnomon on the surface of the earth may be considered as if it is at the center of the earth; (2) the horizon halves the universe; (3) the longest day of the year equals the longest night. Such considerations simplify the problem of projecting a specific part of the path of the sun through a gnomon point onto a shadow-receiving surface. They justify the methods of Vitruvius and Ptolemy described in the Appendix. Both these men recognized that two important parameters sufficiently determined the sun's path observed at any place on earth: geographical latitude and obliquity of the ecliptic. Gnomons, shadows, and perhaps even primitive sundials helped to establish the values of these parameters necessary for the construction of the dials which survived to be catalogued.

According to Eudemus of Rhodes (320 B.C.) quoted by Theon of Smyrna, Oenopides of Chios (ca. 430 B.C.) was the first to observe the obliquity of the ecliptic.[19] Later astronomers determined the value of this obliquity. Eudemus records the value as 1/15 of a circle,[20] and although Eratosthenes (250 B.C.) and Hipparchus (150 B.C.) adopted the better value, 11/83 of a semicircle, it is Eudemus's value that is used in the analemma for the construction of dials described by Vitruvius. Determinations of obliquity no doubt required observations of the solstices and equinoxes, which, as we have indicated, involved γνώμονες.

We know from preserved accounts that the concept of geographical latitude, with values that depend on angular distance from the equator, was first realized from observations of length of daylight and length of a gnomon's equinoctial shadow, both functions of latitude. At the end of the fourth century B.C., the Greek traveler, Pytheas, recorded the ratio of a gnomon to its equinoctial noon shadow for Massalia (Marseilles).[21]

Strabo (2.77) reports that at about the same time, a certain Philo made similar observations at Meroë. In the third century B.C., Eratosthenes combined the available observational data to determine various zones on the earth, within which certain celestial phenomena (height of the pole, longest daylight, gnomon-shadow ratio) did not change appreciably.[22] Hipparchus (150 B.C.) is known to have been concerned with quantitative determinations of latitudes expressed in degrees (1/360 of a circle) north of the equator.[23] His basic observation was to measure shadow length, and he computed ratios of the kind first given by Pytheas. Astronomers and diallers (see the section of the Appendix on Vitruvius) continued to use the gnomon-shadow ratio, with or without converting it into degrees of latitude, to indicate positions on the earth even into the second century A.D., despite Ptolemy's remark in the *Almagest* (2.5) that this ratio cannot be found with certainty.[24]

Very often in Hellenistic geography, terrestrial latitude is expressed as the ratio of longest to shortest daylight. The ancient climata, or parallel geographical zones, were defined in terms of this ratio. Even though variations depending on changes in geographical location may have been observed, it is unlikely that the length of longest daylight, limited by the horizon, was ever directly measured. Hipparchus and other Hellenistic astronomers probably computed lengths of longest daylight from measurements of shadow length using plane trigonometric methods like that described by O. Neugebauer in *The Exact Sciences in Antiquity*, pp. 214-15.

The realization that the appearance of all solar phenomena is affected by changes in latitude apparently did not come as easily to the Romans as it did to the Greeks. Pliny (*Natural History* 7.213) records the transfer of a sundial (*solarium horologium*) from the Greek colony of Catania to Rome in 262 B.C. Romans consulted it with confidence for almost a century before its accuracy was questioned.[25]

We know that the early instruments used to determine dialling parameters functioned like sundials in that they marked solar position by cast shadows. We do not know that these instruments marked more than one solar position per day. Astronomers at the time of Hipparchus must have had access to some means of accurately marking the seasonal hours for they recorded the times of the appearance of celestial phenomena in terms of these ὧραι καιρικαί.[26] From his *Commentary on Aratus* we know that Hipparchus also made use of equinoctial hours in referring to celestial events; and Hellenistic astronomers eventually adopted these ὧραι ἰσημεριναί as basic units of time measurement in their astronomical work, even as they designed sundials to divide the daylight into seasonal hours. We have no evidence that any Hellenistic astronomer-dialler designed a dial which marked equinoctial hours throughout the year. Three dials in the catalogue were designed to emphasize the winter hour as a basic unit for comparative purposes (4001G, 1068G, 1044), but the equinoctial hour is never emphasized.

In the fourth century B.C., members of the Peripatetic School compiled a book of scientific problems which includes questions relating to the theory of gnomonics. Two examples are "Why does the sun produce long shadows when rising and setting, but smaller ones when high in the sky and smallest at midday?" and "Why is it that, although the sun's motion is uniform, the increase and decrease in length of the shadows is not the same in equal periods of time?" (*Problems* 15.9 and 14) It is unlikely that the makers of the dials in the preserved corpus, who were no doubt more interested in technical problems of construction, considered at all the problems of concern to the Peripatetic School. However, we do have indications that men of science in antiquity were aware of theoretical problems relating to gnomonics. By including mathematicians in his list of designers of sundials, Vitruvius emphasizes the relationship between theoretical gnomonics and other exact sciences.

Menelaus's work in spherical trigonometry in the first century A.D. seems to have been motivated by a desire to solve directly problems connected with calculation of length of daylight.[27] Before Menelaus, Greek astronomers solved this and other problems involving the celestial sphere by means of plane trigonometry, considering celestial circles rotated or projected onto a plane.[28] The analemmas of Vitruvius and Ptolemy are examples of the reduction of the problem to graphic solutions in the plane (see Appendix).

Division of the daylight arc requires, and may, as Neugebauer suggests,[29] have inspired, solutions of the classical trisection problem. Neugebauer also sees the sundial as possible motivation for the earliest known definition of conics, represented by the work of Menaechmus in the fourth century B.C. Certainly a plane sundial with perpendicular gnomon always pointing to the culminating sun embodies the conditions which determine a hyperbola in Menaechmus's theory. The case for this direction of influence is, however, weakened by a lack of examples of appropriately adjustable plane dials.[30]

The origins of the art of dialling are hidden in vague literary allusions and suggestive developments in the exact sciences that may be related to it. But the full development and eventual extent of the art is well illustrated by preserved examples. By studying their products, we can evaluate the skill of Graeco-Roman diallers and determine the extent to which they made use of available astronomical and mathematical knowledge. We can estimate the role of the sundial in Graeco-Roman antiquity; and we can suggest possible production methods. The catalogue which forms the major portion of this work is a comprehensive presentation of archeological clues. Each catalogued example provides, in a sense, unique information. However, criteria for evaluation of this information can be grouped according to the type of dial face. The next chapter is devoted to establishing these criteria.

2 Methods of Analysis

In this chapter I analyze, in modern trigonometric terms, the various types of ancient dials. This is neither an attempt at reconstructing the ancient theory nor at discovering how the lines were placed on the dial faces. Rather it is aimed at justifying methods of finding characteristic parameters and important dimensions from measurements made on often poorly preserved dials. Perhaps the most useful piece of information which can be extracted from preserved evidence is the terrestrial latitude for which each dial was intended. For the latitude at which a dial would "work" does not in every case agree with its "known" provenance. In addition, careful analysis of engraved lines on the dial face enables one to theoretically replace missing gnomons.

1. Spherical Sundials

The shadow-receiving surfaces of about eighty-five Greek and Roman spherical sundials are preserved. In most cases the original gnomons do not survive. However, the networks of hour lines and day curves indicate that diallers did not always place gnomon points in the center of the spherical dial face. In some cases, the gnomon point was placed within the spherical surface, but not at its center, or on the spherical surface as in the roofed dials, where the face is in shadow and where the gnomon point is a small hole permitting the passage of sunlight.

a. Hemispherical Dials--Central Gnomon Point

The parameters of a hemispherical sundial having a centrally located gnomon point are easily derived when the dial is compared to the celestial sphere which it reflects. In figure 1 the great circle DBT, parallel to the horizon, marks the top edge of a hemispherical shadow-receiving surface; its center O, is the gnomon point. On a great circle

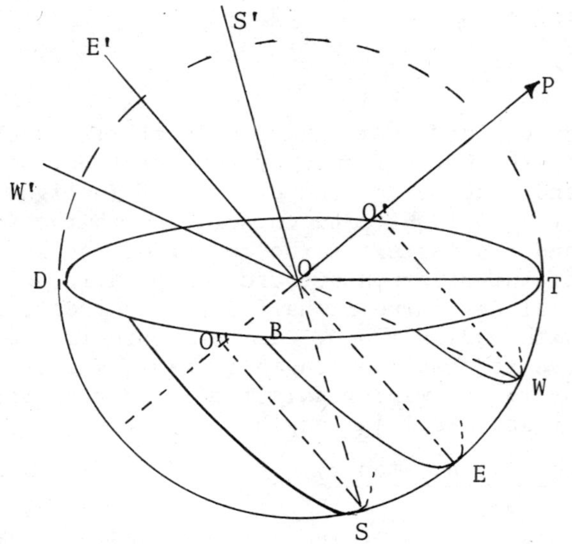

Figure 1

in the meridian plane, W, E, and S mark the shadow of the gnomon point at noon on the winter solstice, the equinoxes, and the summer solstice, respectively. The noon hour line will be the arc of the great circle connecting W, E, and S. The distances in degrees between these three noon hour points on the meridian line are equal, since angle W'OE' = angle E'OS' = ε, where ε is the obliquity of the ecliptic. Also, if r is the radius of the hemisphere and the length of the gnomon,

$$\text{arc WE} = \text{arc ES} = \frac{\varepsilon \cdot \pi \cdot r}{180^\circ} .$$

Hence, on an accurately marked hemispherical dial with central gnomon point the arc lengths WE and ES should be equal. It is a simple matter to select the dials which satisfy this criterion from among the measured catalogue entries. On such dials, the radius or gnomon length, r, can be computed from measured arcs WE or ES:

$$r = \frac{180^\circ \cdot \text{WE}}{\varepsilon \cdot \pi} = \frac{180^\circ \cdot \text{ES}}{\varepsilon \cdot \pi}$$

The arc length TE is a measure of the colatitude, $\bar{\phi}$, intended by the dialler, since

$$\text{angle TOE} = \bar{\phi} = \frac{TE \cdot 180^\circ}{\pi \cdot r} .$$

The day curves on a hemispherical dial with central gnomon point are curves formed by the intersection of the hemisphere and cones generated by rays of sunlight. Since the axis of the sunlight cone coincides with the axis of the earth, and its vertex is the center of the hemisphere, the curves of intersection are arcs of parallel circles. The equinoctial day curve is half a great circle. For a hemisphere with radius r = OT, as in figure 1, the equinoctial day curve is equal in length to $\pi \cdot OT$. The day curve corresponding to the winter solstice is shorter than the equinox curve and equal in length to

$$\frac{WO' \cdot \pi \cdot \cos^{-1}(\tan\phi \cdot \tan\varepsilon)}{90^\circ}$$

where $WO' = OT \cdot \cos\varepsilon$, ϕ is the latitude and $\cos^{-1}(\tan\phi \cdot \tan\varepsilon)$ is half the daylight arc. It happens that for the relevant latitudes (that is, above 19°), the day curve corresponding to the summer solstice is longer than the equinoctial day curve and equal in length to

$$\frac{SO'' \cdot \pi \cdot \cos^{-1}(-\tan\phi \cdot \tan\varepsilon)}{90^\circ}$$

where $SO'' = OT \cdot \cos\varepsilon$ and $\cos^{-1}(-\tan\phi \cdot \tan\varepsilon)$ is again half the daylight arc.

b. Cut Spherical Dials--Central Gnomon Point

The gnomon point's shadow will never fall on the part of a hemispherical surface south of the summer solstice curve. On most known examples of spherical dials with centrally located gnomon, that southern portion has been cut away so that the front edge of the dial lies in a plane parallel to, or sometimes coinciding with, the summer solstice curve. The angle at the upper edges of the dial is then equal to the angle between solstice curve and horizon, or $\bar{\phi}$. Such a cut spherical dial, hollowed from an angular block of stone, is illustrated in profile in figure 2. It is obvious that a direct measurement of $\bar{\phi}$ and, if the important corner has been damaged, of ϕ can be taken on preserved dials cut in this manner.[1] In figure 3, which illustrates the horizontal

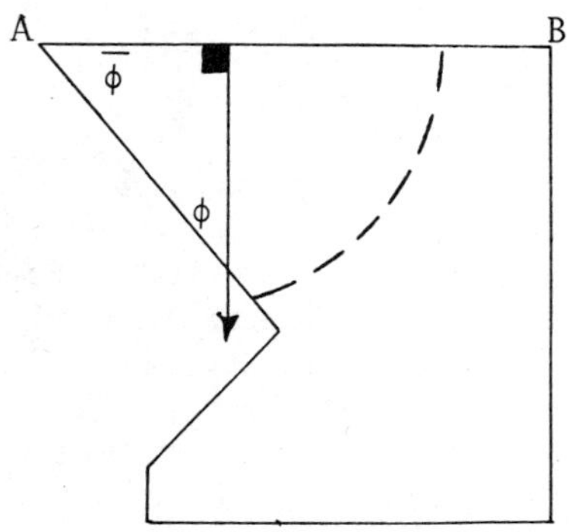

Figure 2

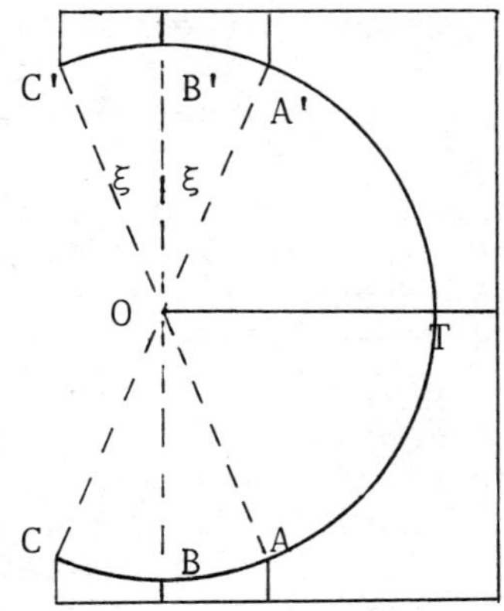

Figure 3

section of a cut spherical dial, A, B, C and A', B', C' mark the points of intersection of the solstitial and equinoctial day curves and the horizon. The arc distances AB and BC are both equal to $(\pi \cdot OT \cdot \xi)/180°$, where ξ is the ortive amplitude of the sun:

$$\sin \xi = \sin \varepsilon / \sin \phi.$$

The shadow of a gnomon point in the center of a spherical surface will travel one twelfth of its daily path in each of the ancient seasonal hours. If each possible daily path is divided into twelve equal parts, the points of division will, throughout the year, trace out thirteen hour lines, of which eastern and western horizon are two, and the meridian a third. The seasonal hour lines, with the exception of the meridian and horizon, are not simple curves. T. S. Davies[2] was the first to analyze the equations of these curves referred to great circle coordinates originating at the point of intersection of equator and meridian:

$$\tan \delta = \tan \overline{\phi} \cdot \cos n \cdot \sigma$$

where δ is the solar declination, $\overline{\phi}$ is the colatitude, σ is the number of degrees of the arc of its path traveled by the sun since sunrise up to noon; further, $n = 6/m$, where m is the number of seasonal hours since sunrise up to noon.

He concluded that the curves have no characteristic in common with circles. They are curves of intersection of the spherical shadow-receiving surface and an "undulating cone" with vertex at the gnomon point and equation

$$\frac{z}{\sqrt{(x^2 + y^2)}} = \tan \overline{\phi} \cdot \cos \left[n \cdot \cos^{-1}\left(\frac{x}{\sqrt{(x^2 + y^2)}} \right) \right]$$

referred to orthogonal axes in the equator and meridian planes with origin at the gnomon point.

In spite of their noncircular nature, for latitudes below 45° the seasonal hour lines between meridian and horizon are very closely approximated by the great circles which *do* pass through corresponding seasonal hour points on solstitial and equinoctial curves.[3] The engraved hour lines on preserved spherical sundials appear to be such great circle approximations. The deviation of the hour lines from great circles cannot even be detected on the few dials where more than three day curves have been divided (see 1068G). All of the known spherical dials were made to be used at latitudes below 45°. We have no way of knowing whether the ancients were aware of the very elongated "S" shape of accurate hour lines.

The segment between hour points, ℓ_δ, on any day curve is characteristic of that day curve, for

$$12 \cdot \ell_\delta = \frac{\pi \cdot r \cdot \cos \delta \cdot \cos^{-1}(-\tan \phi \cdot \tan \delta)}{90°}.$$

At the equinox, $\delta = 0$ and $12 \cdot \ell_0 = \pi \cdot r$. Thus, when it is possible to measure a segment ℓ_0 on a preserved spherical dial with central gnomon, it is in turn possible to determine the radius of the sphere,

$$r = OT = \frac{12 \cdot \ell_0}{\pi} .$$

Where it is also possible to measure a segment of solstice curve, one can estimate the latitude assumed by the dial maker since

$$\tan \phi = -\cos \delta \cos \left[\frac{90° \cdot 12 \cdot \ell_\delta}{\pi \cdot OT \cdot \cos \delta} \right].$$

Internal consistency demands that values for ϕ (and $\overline{\phi}$), ε, and OT calculated from measurements taken on the day curves agree with directly measured values of ϕ or $\overline{\phi}$ (see figure 2) or values computed from measurements taken on the meridian or horizon lines. For spherical dials with central gnomon point, the simple straight gnomon length must, of course, be equal to r.

The popularity of spherical surface dials with oblique fronts in a plane parallel to the plane of the equator is likely due to the simplicity with which such dials could be constructed. On a properly cut sphere with central gnomon point all curves equidistant from the front edge will be parallels of declination. The divided day curves engraved on the dial face need *not* represent the solstices and equinox. If the dial is to work as an accurate calendar as well as a clock, then the parallels engraved on the spherical shadow-receiving surface should, of course, correspond to parallels of the sun's declination, with the equinox being a great circle arc on the sphere. However, if the dial is to be used primarily to mark the hours, then the dial maker need only divide two or perhaps three sufficiently separate parallel curves and connect corresponding hour points with approximate circular arcs on the sphere.

Several examples of oblique front spherical dials found at Roman sites were constructed using equally divided parallel circles which do not designate solstices or equinox. The dials do mark the seasonal hours well, although their calendrical function is unclear. Measurements taken on dials from Pompeii, Herculaneum, Ostia, and Rome[4] show that the hour lines were constructed using arbitrary parallel circles by makers who were not particularly concerned with exact seasonal markings. Photographs of Roman dials from North Africa and measurements of a Roman dial found at Cannstadt show hour lines that extend between two equally divided engraved day curves probably corresponding to the solstices, and indicating, I think, a lessening emphasis on the calendrical role of the dial. Continuing in this tradition, two dials from Sicily have been marked with hour lines only, which stretch from an equally divided oblique front edge (parallel to the plane of the equator) to the gnomon groove on the meridian.

The intended latitude of cut spherical dials which have been graduated without reference to specific solar declination can only be determined from measurements of angle α or β (see figure 4). The radius of the receiving surface and hence the gnomon length can be found by determining the center of a trace of the edge of the horizontal top surface of the dial. Once these parameters have been established, it is possible to determine just which parallel circles

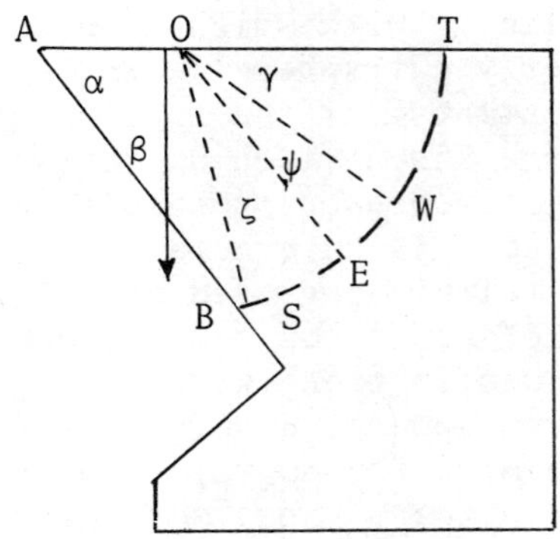

Figure 4

have been divided to form the hour lines and to use this information to test internal consistency.[5]

c. Quarter Spherical Dials--Central Gnomon Point

There are a few spherical dials preserved from Roman antiquity to which Vitruvius's descriptive phrase, *hemicyclium excavatum ex quadrato ad enclimaque succisum*, cannot be applied. These dials are *not* cut according to the latitude of their location, but consist of a quarter spherical surface with horizontal top edge, and front edge parallel to the plane of the prime vertical.[6] "Day curves" on these quarter spheres are circular arcs lying in planes parallel to the front edge. The best example of this type of dial is now in the National Archeological Museum in Bucharest (1048G). Figure 5 gives a vertical cross section at the meridian. The "day curves" lie in the planes SS', EE', and WW'. Arcs connecting the corresponding points of equal duodecimal division of these curves form the hour lines. At noon the hour points W, E, S coincide with points on parallel circles traced by the shadow of a central gnomon point at the solstices and equinoxes. Given these characteristics of the dial's network of hour lines and day curves, it is

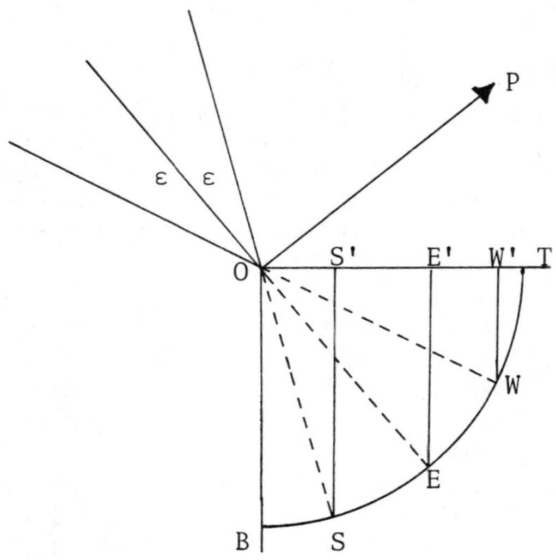

Figure 5

possible to determine the latitude for which it was constructed and the length of its horizontal gnomon. In fact these values can be computed in more than one way and the results compared as a test of internal consistency.[7]

On some quarter spherical dials with central gnomon point, only a semicircular equinoctial day curve is divided by hour lines radiating from the gnomon hole. Once the radius of the sphere has been determined by finding the center of a trace of the curve of front or top edge, one can use it and the measured equinoctial arc length between hour lines to calculate the intended latitude. Clearly these dials do not measure the seasonal hours accurately. They were products of relatively unsophisticated diallers.

d. Hemispherical Dials--Noncentral Gnomon Point

In 1907 a unique hemispherical dial was found in the ruins of ancient Pergamon.[8] Two separate networks of hour lines and day curves had been engraved on the dial's face, corresponding to the shadows cast by two distinct horizontal gnomons whose endpoints were in the meridian plane, but not at the center of the hemisphere. The designer of the Pergamon dial understood that the day curves traced by a noncentral gnomon point on a spherical face are not parallel circles, and that the hour points, while symmetric about the meridian, do not equally divide these day curves. He probably determined the distinctive pattern of day curves and hour lines by using projection methods which do not require an understanding of conic sections or spherical geometry. The design offers certain economic advantages in that a single-gnomon model requires less material than a dial with central gnomon point. It is quite possible that other dials of the Pergamon type with one or two faces may yet be recovered.

On the Pergamon dial, the positions of the gnomon points are determined by the condition that the winter solstice curve traced by the shadow of the point of the gnomon attached to the north-facing surface of the hemisphere coincides with the summer solstice curve traced by the shadow of the point of the gnomon attached to the south-facing surface of the hemisphere. Day curves on this dial are parts of higher order curves resembling arcs of ellipses, which can be seen by recognizing these curves as the result of noncentral intersections of sunlight cones with a spherical surface as in figure 6. The vertex of the sun-

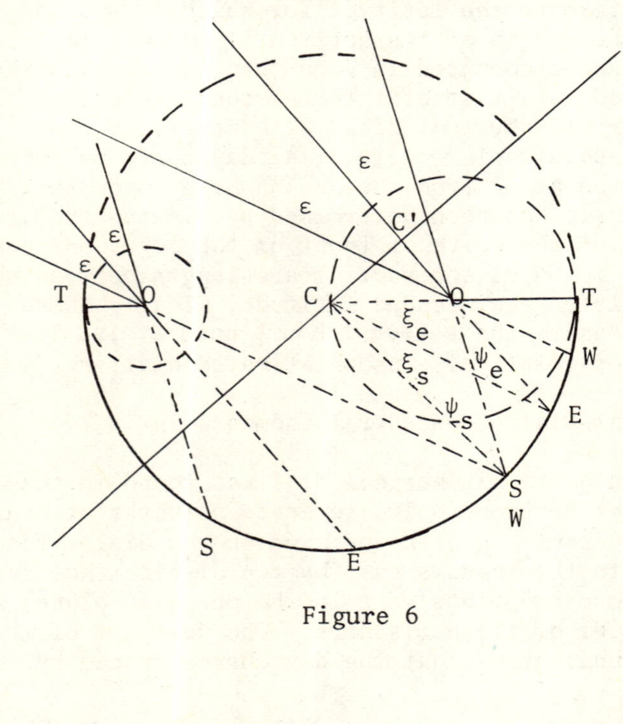

Figure 6

light cone, O, does not coincide with the center of the intercepting spherical surface, C. The day curves at solstice are projected onto parts of parabolas in the meridian plane and at the equinoxes they become an arc of a circle (not a semicircle).[9] The departure of the solstitial curves from circles is most apparent on the north-facing portion of the hemisphere.

Figure 7 illustrates the horizontal section of the dial

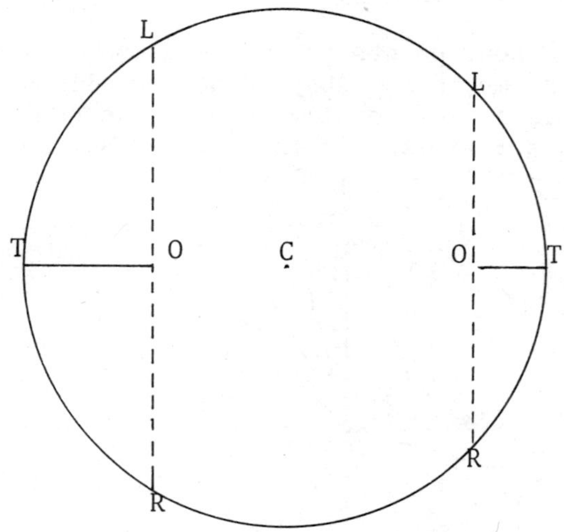

Figure 7

where L and R mark the intersection of the equinoctial day curve with the top edge and OT is the gnomon length g. Both the distance LR (equal to $2\sqrt{(2r - g)g}$) and the distance TT (equal to 2r) can be measured, leading to determinations of g and r. Once the radius, r, of the hemispherical surface and the gnomon length, g, have been determined it is possible to find the latitude for which the dial was constructed by using measurements taken along the meridian line (see figure 6). The distances WE and ES are not equal. In sector TCE

$$\xi_e = \frac{180° \cdot TE}{r \cdot \pi} .$$

Given CO, ξ_e, and CE = r in triangle COE (figure 6)

$$OE = \sqrt{(CO^2 + CE^2 - 2CO \cdot CE \cdot \cos \xi_e)}$$

and

$$\psi_e = \sin^{-1}\left[\frac{CO \cdot \sin \xi_e}{OE}\right]$$

then

$$\text{angle } TOE = \overline{\phi} = \xi_e + \psi_e.$$

Although the hour points on the Pergamon dial are symmetric about the meridian, they do not equally divide the day curves. This is most easily seen at the equinox. In figure 8, EOL is the plane of the path of sun and shadow,

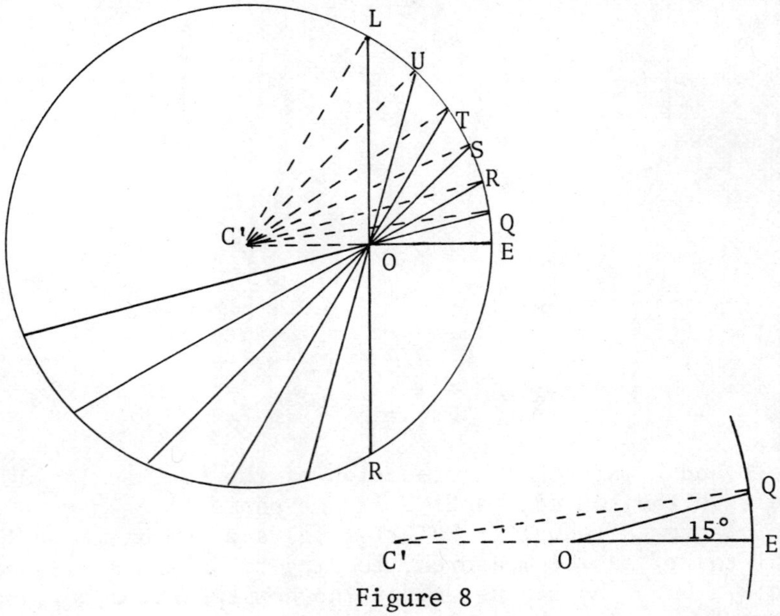

Figure 8

i.e., the equator. For angle EOQ = angle QOR = 15°, $EC' = CE \cdot \cos \psi_e = C'L$, $C'O = \sqrt{(C'L^2 - LO^2)}$, and $OE = EC' - C'O$. Triangle OC'Q can be solved for angle C'QO where

$$C'QO = \sin^{-1}\left[\frac{C'O(\sin QOC')}{C'Q}\right]$$

and

$$QC'O = 15° - C'QO.$$

Succeeding angles at C' will not be multiples of angle QC'O and hence the arc lengths EQ, QR, etc., will not be equal although they may approximate each other. At the solstices where the divided day curves are of higher order, the distances between hour points are even more obviously unequal.

e. Roofed Spherical Dials

Twenty-four spherical sundials have been found with a network of hour lines and day curves which corresponds to a "gnomon point" on the spherical surface. Figure 9 shows

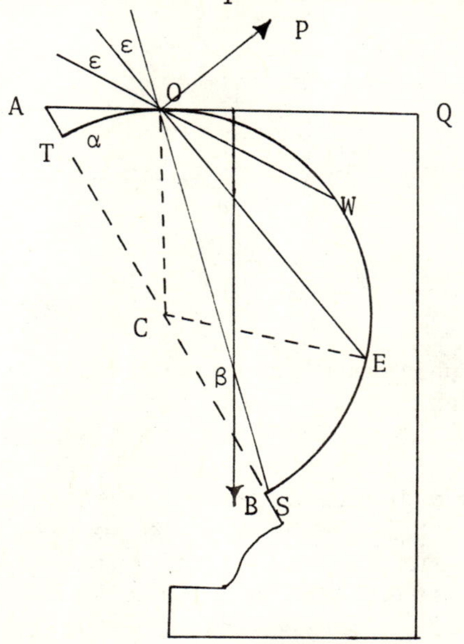

Figure 9

the vertical section of a well-preserved example of this type of dial; O marks the position of the "gnomon point." The flat (or slightly rounded) top of the dial, AQ, forms a kind of ceiling, intercepting the rays of sunlight and always leaving the concave spherical surface at least partially shadowed. If the ceiling is not complete, but cut away near the meridian and notched at the highest point of the underlying sphere (as in figure 10), the position of the sun will be marked by the notch at the border of sunlight and shadow on the sphere. If the ceiling is complete and pierced at the highest point of the underlying sphere (as in figure 11), then the position of the sun will be

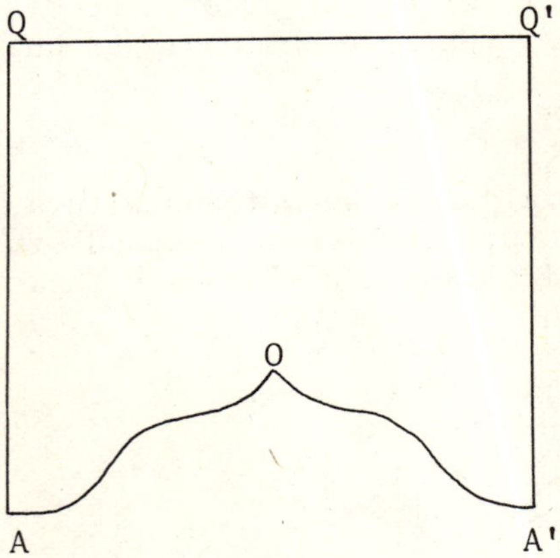

Figure 10

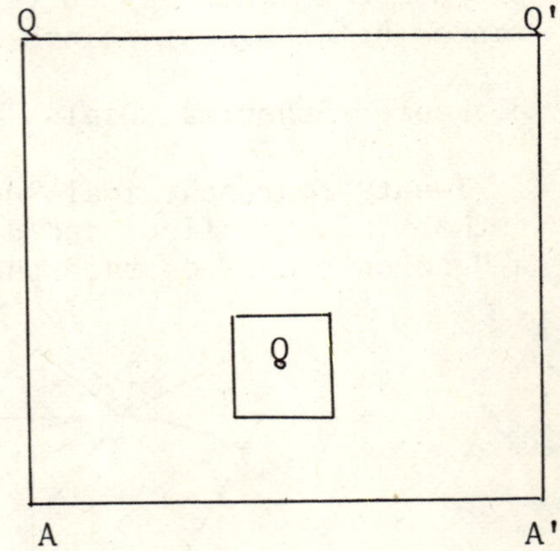

Figure 11

traced by rays of sunlight which shine through the pierced hole and form a bright image of the sun on the shadowed surface of the dial. At least four preserved examples of this spherical variant dial have a complete pierced ceiling. Three of these show evidence of having been fitted with a thin metal plate which could be made to conform more closely to the spherical surface and which could be pierced more accurately than the stone. The ceilings of the twenty or so remaining variants of this type are unfortunately too damaged to serve as evidence of the use of either a notched edge or a small hole as gnomon point.

In theory, a day curve on a roofed spherical dial is the curve of intersection of a cone and a sphere where the vertex of the cone lies on the spherical surface at the hole or notch which acts as gnomon point. Figure 9 illustrates the conditions which prevail at noon on the solstices and equinoxes. It can be shown that the sun's path, traced on the dial face at these times of year, will, when projected onto the plane of the equator, be a conchoid curve, Pascal's limaçon.[10] The configuration of day curves on the spherical surface is basically a three dimensional version of this conchoid curve.

On a roofed spherical dial, the distances between solstitial and equinoctial lines at noon should be equal. As can be seen in figure 9 they subtend angles at the center of the spherical surface equal to twice the obliquity of the ecliptic. The radius of the sphere, r, is then equal to $(180^\circ \cdot WE)/(2 \cdot \varepsilon \cdot \pi)$ or $(180^\circ \cdot ES)/(2 \cdot \varepsilon \cdot \pi)$ where arcs WE and ES can be measured on the dial face. The distance OE from gnomon to equinoctial noon mark is, in effect, the diameter of the circular equinoctial day curve. This circle is illustrated in figure 12, divided, as it should be on the dial,

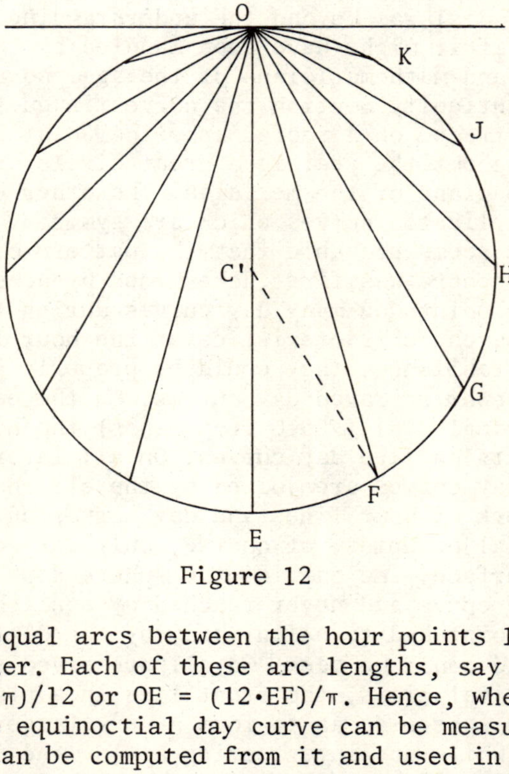

Figure 12

into twelve equal arcs between the hour points 15° distant from each other. Each of these arc lengths, say EF, is equal to $(OE \cdot \pi)/12$ or $OE = (12 \cdot EF)/\pi$. Hence, where an arc length on the equinoctial day curve can be measured, the diameter OE can be computed from it and used in turn to estimate a value for the intended latitude. In triangle COE (figure 9)

$$\frac{\sin 2\overline{\phi}}{\sin \phi} = \frac{OE}{r}$$

or

$$\cos \phi = \frac{OE}{2 \cdot r} .$$

Hour points on the irregular day curves which are the

path of the solar image on the spherical surface at the solstices are not equidistant from each other. Their positions, if determined by some method other than trial and error, were very likely found by means of projection. Albert Rehm proposes the use of projection methods in the division of the day curves on the hemispherical dial found at Pergamon.[11] It seems most likely that the higher order curves of a roofed spherical dial were divided in a similar manner. I am sure that a theoretical determination of the distances between hour points on the solstitial curves of this type of dial was beyond the understanding of most dial makers even if it might have been handled by the celebrated astronomers and mathematicians of the age. No surviving works from antiquity mention the curves in question.

The line traced on a roofed spherical dial by noon hour points throughout the year is a great circle of the sphere lying in the plane of the meridian. The other hour lines are more complicated curves which are symmetric about the meridian. It seems probable that the pattern of hour lines on a roofed sphere was first determined by actually connecting hour points on many day curves during the year. Once the apparent characteristics of the hour lines had been thus established, they could be properly placed on dials with fewer engraved day curves. On the earliest known roofed spherical dial (7001G from Tenos) the hour lines connect points on five day curves. On all later examples only three day curves are joined by the eleven hour lines.

The network of hour lines and day curves on a roofed spherical dial occupies, of course, only part of the spherical surface. The part of the sphere south of the summer solstice curve and never reached by sunlight through the gnomon point had to be cut away by the dialler so the engraved grid could be seen. On all preserved examples of roofed spherical dials, the front lies in a plane which intersects the sphere so as to form a very large front circular opening while not intersecting any of the day curves (see figure 9). On some dials, the angle β between front plane and vertical plane is equal to ϕ, but on the majority of dials β is less than ϕ. In fact, any plane passing perpendicularly through a line joining the center of the sphere to a point on the meridian south of the summer solstice will serve the purpose.

Not all the measurements included in the catalogued entries of roofed spherical dials conform to the requirements of theory. On some dials the arc lengths WE and ES differ slightly, or the twelve parts of the equinoctial curve are not exactly equal. These variations weaken tests for internal consistency. They do, however, give valuable information about the construction of this type of dial. The diallers were either unaware of or unable to comply with the theoretical requirements of the design.

f. The Spherical Globe

In 1939 Carl W. Blegen published a brief description of a global "sundial" which he found among the ruins of ancient Prosymna (catalogue number 7002G). He attributed the functional inscription to the second century B.C., concluding that the sphere was made at that time. Blegen was justifiably puzzled by the network of lines on his globe. The sphere remains a uniquely curious example of the work of an ingenious dialler.

Of the three networks on the surface of the globe, that which can be most easily interpreted consists of a series of seven parallel arcs resembling the day curves on a concave spherical sundial. The location of these seven arcs is shown in figure 13, a vertical cross section of the globe

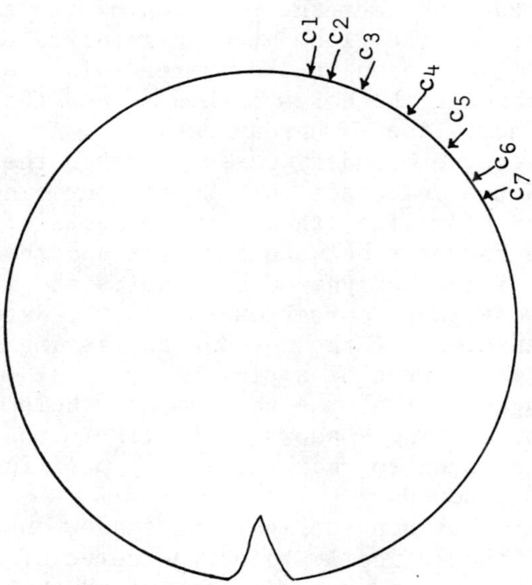

Figure 13

oriented with the conical depression in its surface at its lowest point. The area between curves c_1 and c_2 on the Prosymna globe is marked with the names of the zodiacal signs which immediately precede and follow the summer solstice. Similarly, the area between curves c_6 and c_7 is marked with the names of the zodiacal signs which immediately precede and follow the winter solstice. The placement of other zodiacal names within the network suggests that the parallel curves are associated with the zodiacal divisions of the sun's yearly path, with curve c_4 lying in the plane of the celestial equator. The seven zodiacally related day curves are cut by thirteen hour lines, each line pierced somewhere along its length by a small shallow hole designated by a Greek-letter numeral.

The Greek-letter designations, the placement of the shallow holes, and the nature of the engraved network of lines support the conclusion that in the course of its movement across the surface of the sphere, the *boundary between light and shadow* marks the seasonal hours. When in use, the globe would have been mounted on a pole with the hole marked "IB" at its highest point. The network of day curves and hour lines would have faced south with the holes marked ϛ12 due east and west. To appreciate the modus operandi of the dial one can consider the position of the sunlight-shadow boundary on the globe at noon in different seasons (see figure 14). At noon on the summer solstice the boundary passes through the holes marked ϛ and the intersection between hour line VI and day curve c_1. At noon on the winter solstice the boundary passes through the holes marked ϛ and the intersection between hour line VI and day curve c_7. Each hour line, then, is the locus of points generated by the boundary between sunlight and shadow as it falls on one of the designated hour holes and intersects the zodiacal day curve corresponding to the date. An hour line is the envelope of the boundaries passing through an hour hole in the course of a year.

Blegen suggested that the shallow hour holes once held brass pins for casting shadows. More likely the holes held a movable brass stud for marking timely positions of the sunlight-shadow boundary.

The boundary between sunlight and shadow on a globe can be considered theoretically to be the curve of tangency between the globe and an encompassing cylinder of equal diameter. The orientation of the cylinder relative to the fixed globe changes with solar position.

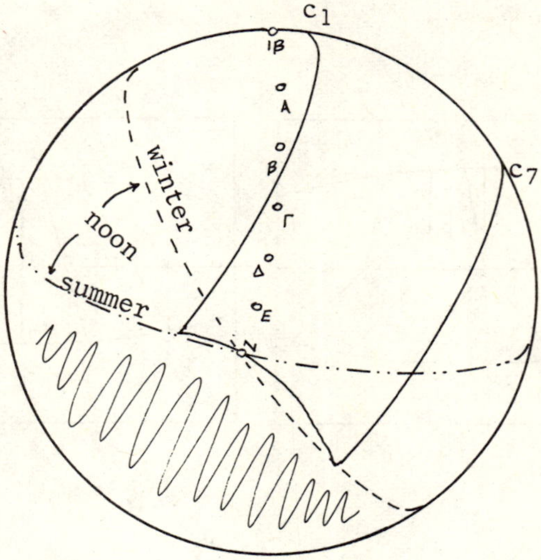

Figure 14

The designer of the Prosymna globe included a north-facing network of lines which also utilizes the sunlight-shadow boundary as it passes through the hour holes used with the south face. Again intersection between appropriately marked boundary and day curve marked the hour. The resulting network is appropriately similar to the network on a roofed spherical dial, for the shadow boundaries which determine the one are orthogonal to the rays of sunlight which cast an image of the sun on the other.

A third network of lines stretches between the east and west hour holes across a shallow hole in the globe's meridian. I interpret it as a secondary south-facing map of the moving boundary between sunlight and shadow. The three great semicircles which form part of the network are divided by points which would lie along the hourly positions of the boundary at winter solstice (lower semicircle), equinox (middle semicircle), and summer solstice (upper semicircle). One might have expected the dial's maker to connect analogous divisions on each curve to form hour lines, but he has apparently confused his connections, creating a group of lines more decorative than they are useful. Figure 15 shows the error and a proposed correction.

If this account of the spherical globe has seemed par-

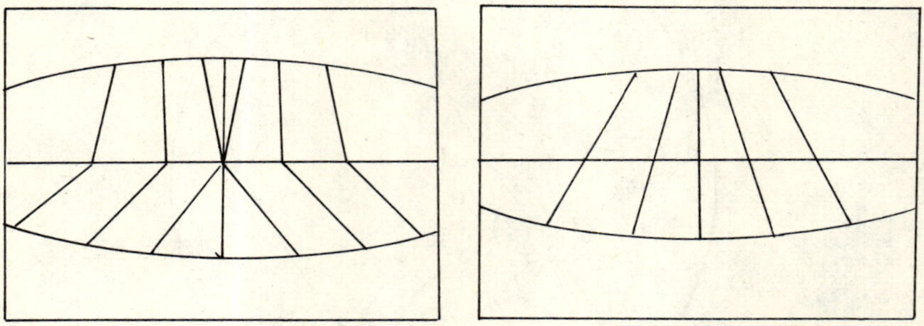

Figure 15

ticularly specific, it is because only one example is known. The design is not mentioned in ancient accounts (see chapter 3), but it may be part of the celestial globe tradition begun, according to Cicero,[13] by Thales of Miletus "and later marked by Eudoxus."

2. Conical Sundials

A very large number of sundials found at Greek and Roman sites around the Mediterranean have a face which is part of the surface of a cone with its axis parallel to the axis of the earth. The conical face is bounded at the top by a plane parallel to the horizon and in front by a plane parallel to the equator. Hour lines and day curves on these dials form a pattern which varies with the position of the vertex of the conical face and also with the position of the gnomon point. The vertex of the conical surface may lie above, below, or on the horizontal top surface of the dial.

a. Vertex Above the Horizontal Top Surface--South-facing Conical Dials

At least ninety sundials from Graeco-Roman antiquity have a conical face which is distinguished by the following characteristics: first, the curve at the front edge of the dial face is circular; second, the engraved day curves on the dial face are circular arcs parallel to the front edge; third, the hour points equally divide the day curves; fourth, the noon line is straight and coincides with a generator of the conical surface; fifth, the distance between winter solstice curve and equinox along the noon line is less than the distance between equinox and summer solstice curve. These characteristics indicate that the dial face must be part of a right circular cone with vertex above the horizon and axis coinciding with the common axis of the sunlight cones. As vertex of the sunlight cones, the gnomon point must then be on this common axis. Figure 16 shows such a

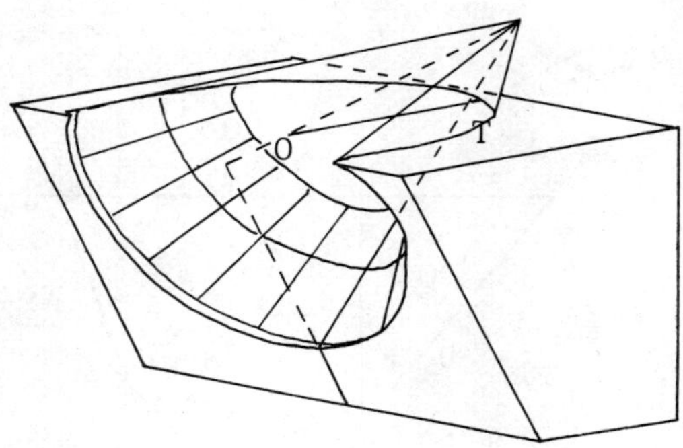

Figure 16

conical dial face with gnomon OT. All preserved conical dials of this type were designed to face south. On four of these (3001G, 3023G, 3024G, 3064G) preserved gnomons extend horizontally from the top surface in the meridian plane. The shape and location of the sockets on other dials indicate that the gnomon was always fixed in this position.

In theory the hour lines on a conical dial are not simple curves, as T. S. Davies has shown.[14] In practice the lines have been approximated with acceptable accuracy by the arcs of conic sections which connect corresponding hour points on the solstitial and equinoctial day curves. On one dial (3059G) only solstitial hour points are connected by approximate hour lines. On a few other dials the hour lines radiate from the gnomon hole to equally divide the circular front edge without crossing any engraved hour lines.

The parameters involved in the construction of a conical sundial include: the terrestrial latitude of the dial's intended location, ϕ; the angle ω between the axis and a generator of the dial surface; the length of the gnomon, g. By analyzing a conical dial's network of hour lines and day curves, it is possible to find these determining parameters.

The angle ω between the axis and a generator of the conical face can be computed from measurements of the arc length between two consecutive hour points on the equinox, ℓ_E, and the linear distance between noon points on the equinox and the solstices, WE and ES. In figure 17, the meridian

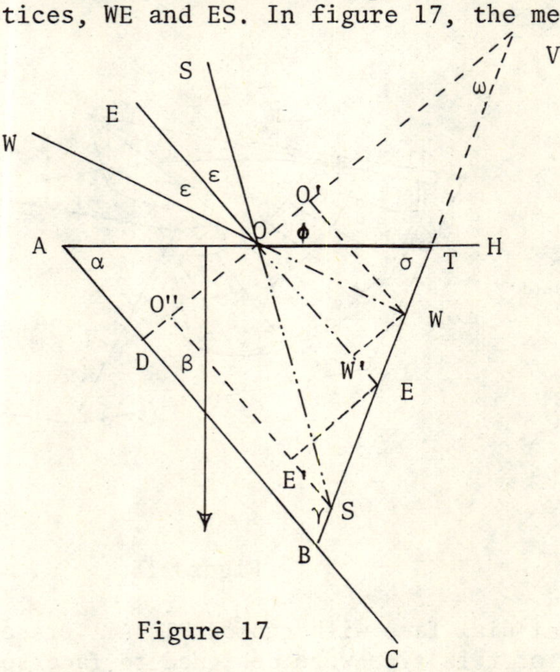

Figure 17

section of a south-facing conical dial, OE is the radius of the equinoctial day curve; $OE = (12 \cdot \ell_E)/\pi$. In triangle EOW, angle EOW = ε, angle OEW = $90° - \omega$, angle EWO = $90° + \omega - \varepsilon$, and

$$\frac{\sin EOW}{WE} = \frac{\sin EWO}{OE} \quad \text{or} \quad \frac{\sin \varepsilon}{WE} = \frac{\sin(90° + \omega - \varepsilon)}{OE}.$$

Similarly, in triangle SOE

$$\frac{\sin SOE}{SE} = \frac{\sin OSE}{OE} \quad \text{or} \quad \frac{\sin \varepsilon}{SE} = \frac{\sin (90° - \omega - \varepsilon)}{OE}.$$

Combining the above equations,

$$\tan \omega = \frac{ES - WE}{ES + WE} \cdot \cot \varepsilon.$$

In theory, the computed value of ω should equal angle TBC - 90° or 90° - γ (see figure 17). The accuracy of a conical dial can be tested by comparing the value of ω determined from measurements of distances WE and ES with the value of ω determined from measurements of angle TBC or γ.

If, in addition to the distances required for determining the angular opening of the cone, one can also measure the distance TW along the meridian, it is possible to find the intended latitude of a conical dial. In the right triangle VOE (figure 17), $VO = OE \cdot \cot \omega$, $VE = OE/\sin \omega$. $VT = VE - TE$ and in triangle VOT,

$$TO^2 = g^2 = VT^2 + VO^2 - 2\ (VT \cdot VO) \cos \omega.$$

Also in triangle VOT,

$$\sin VOT/VT = \sin OVT/TO \text{ or } \sin \phi/VT = \sin \omega/g \quad \text{and}$$

$$\phi = \sin^{-1} (VT \cdot \sin \omega)/g \ .$$

The intended latitude of a conical dial can also be determined from measurements of WE, ES, ℓ_E, and the arc length between two hour points on either solstice day curve, ℓ_W (winter) or ℓ_S (summer).[15] The arc length of one half of the winter solstice curve, $6\ell_W$, is related to the angle of half daylight, τ_W, by the equation

$$\tau_W = \frac{6 \cdot \ell_W \cdot 180°}{O'W} \tag{2.1}$$

where O'W is the radius of the winter solstice curve (see figure 18, which shows this curve in the equator plane). In triangles OWO' and OSO" in figure 17,

$$\tan \varepsilon = OO'/O'W = OO''/O''S \quad \text{or} \quad OO'/OO'' = O'W/O''S.$$

Since OO' = WW' and OO" = EE', OO'/OO" = WW'/EE' = O'W/O"S. In similar triangles WW'E and EE'S

$$\frac{WW'}{EE'} = \frac{WE}{ES} = \frac{W'E}{E'S} \quad \text{or} \quad \frac{O'W}{O''S} = \frac{WE}{ES}. \tag{2.2}$$

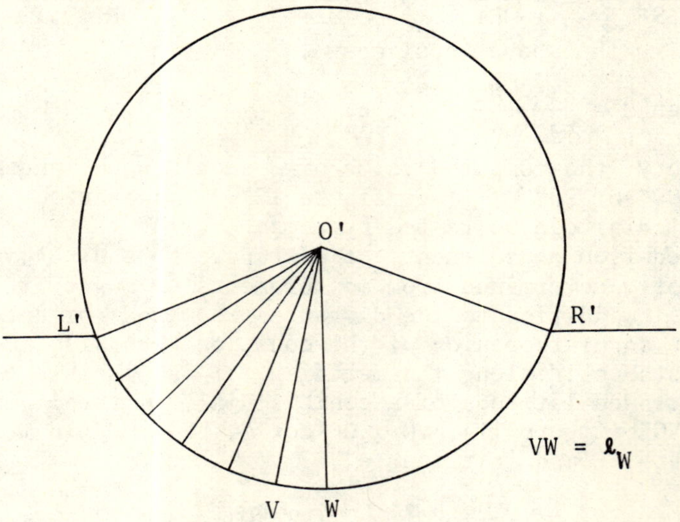

Figure 18

Since O'W = OW' and OE = O"E', W'E = OE - O'W and E'S = O"S - OE so that

$$\frac{WE}{ES} = \frac{OE - O'W}{O''S - OE}\,. \qquad (2.3)$$

From (2.2) and (2.3)

$$\frac{WE}{ES} = \frac{OE - O'W}{[(O'W \cdot ES)/WE] - OE} \text{ or } O'W = \frac{(ES + WE) \cdot OE}{2\,ES}\,.$$

For OE = $(12 \cdot \ell_E)/\pi$,

$$O'W = \frac{(ES + WE) \cdot 12 \cdot \ell_E}{2 \cdot ES \cdot \pi}\,. \qquad (2.4)$$

Substituting (2.4) in (2.1)

$$\tau_W = \frac{180° \cdot ES \cdot \ell_W}{(ES + WE) \cdot \ell_E}\,.$$

According to the modern trigonometric formula, half daylight in degrees at winter solstice is τ_W where $\cos \tau_W = \tan \phi \cdot \tan \varepsilon$. Substituting the value of τ_W computed from the arc length of the summer solstice curve on the conical face:

$$\tan \phi = \cot \varepsilon \cdot \cos \left[\frac{180° \cdot ES \cdot \ell_W}{(ES + WE) \cdot \ell_E}\right]. \tag{2.5}$$

Half daylight at summer solstice can be determined from the formula $\cos \tau_S = -(\tan \phi \cdot \tan \varepsilon)$. Substituting the value of τ_S computed from the arc length of the summer solstice curve on the conical face:

$$\tan \phi = -\cot \varepsilon \cdot \cos \left[\frac{180° \cdot WE \cdot \ell_S}{(ES + WE) \cdot \ell_E}\right]. \tag{2.6}$$

The angle α (angle HAB in figure 17) between the horizontal top surface of a conical dial and its oblique front surface is a measure of the colatitude intended by the dial maker. The angle σ (angle OTE in figure 17) is a measure of the sum of the latitude and the angular opening of the cone. The relative values of σ and ω determine the type of conic section formed by the horizontal top edge of the conical dial: for $\sigma = 2\omega$, the curve at the top edge is a parabola; for $\sigma > 2\omega$, the curve is part of an ellipse; for $\sigma < 2\omega$, the curve is part of one branch of a hyperbola.

b. Vertex Below the Horizontal Top Surface--North-facing Conical Dials

Three ancient sundials have two conical surfaces each, one facing south, as in the ones just described, and an additional face towards the north (see dials 3050G, 3107G, and 7003G). The characteristic pattern of hour lines and day curves on the north face is quite similar to that found on the south face except that the positions of the solstitial curves are reversed. The equally divided circular day curves parallel to the circular front edge again indicate that the face is a segment of a right circular cone with axis coinciding with the axis of the sunlight cones. However, for each of these north faces, the vertex of the segmented cone lies below the horizontal top surface of the dial. Figure 19 shows the meridian section of a north conical face. The gnomon point is at O on the axis. Note that the vertex of the cone, V, is carved into the stone.

The parameters of a conical dial with meridian section as shown in figure 19, i.e., ϕ, ω, and g, can be determined

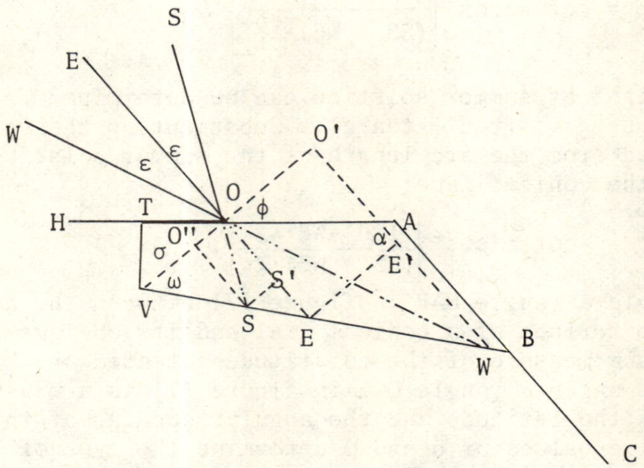

Figure 19

from measurements taken on the network of hour lines and day curves using methods analogous to those just described. In particular, measurements of SE and WE can be combined to find ω since

$$\tan \omega = \frac{WE - ES}{WE + ES} \cdot \cot \varepsilon . \tag{2.7}$$

If the distance between two consecutive hour points on the equinox, ℓ_E, and on the winter solstice, ℓ_W, or summer solstice, ℓ_S, can be measured in addition to SE and WE, then the intended latitude can be determined using formulas (2.5) and (2.6).

The distance along the axis of the conical face from the vertex to the gnomon point, VO in figure 19, can be found by solving the right triangle VOE. OE is the radius of the equinoctial day curve and is equal to $(12 \cdot \ell_E)/\pi$. The angle ω can be found from equation (2.7) and $VO = OE \cdot \cot \omega$. If VT can be measured accurately, the gnomon length $TO = g$ can be found by solving triangle VOT by formula (2.4).

The computed parameters of a north-facing conical dial should agree with angular measurements made on the dial face. For example, the angle α between the horizontal top surface and the surface parallel to the equator should equal 90° + ϕ. The angle σ (angle VTO in figure 19) should be 180°-(ϕ +ω). As was the case for conical dials with vertex above the horizon, the sizes of angles σ and ω determine the character of the curve at the top edge of the dial: where $\sigma = 2\omega$, the curve is a parabola; where $\sigma < 2\omega$, the curve is elliptical; where $\sigma > 2\omega$ the curve is hyperbolic.

It is theoretically possible to design south-facing conical dials with vertex below the horizon (see meridian section in figure 20); however, no Greek or Roman examples of this variant have been found.

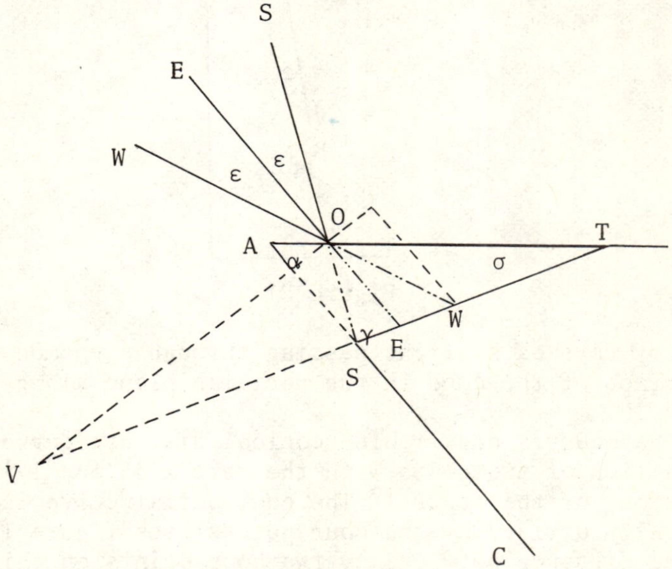

Figure 20

c. Vertex On the Horizontal Top Surface--Roofed Conical Dials

A unique example of a roofed conical dial, analogous to the roofed spherical dials discussed earlier, is in the Basmahane Archeological Museum in Izmir, Turkey (3109G). Figure 21 shows the meridian section of this dial with suggested restoration beyond the broken edge at top front. The vertex, V, of the conical surface is on the horizontal top surface of the dial. The dial face is a complete right circular cone with axis parallel to the axis of the earth and with generator VA parallel to the horizon. The hours are marked on the conical surface by a bright image of the sun

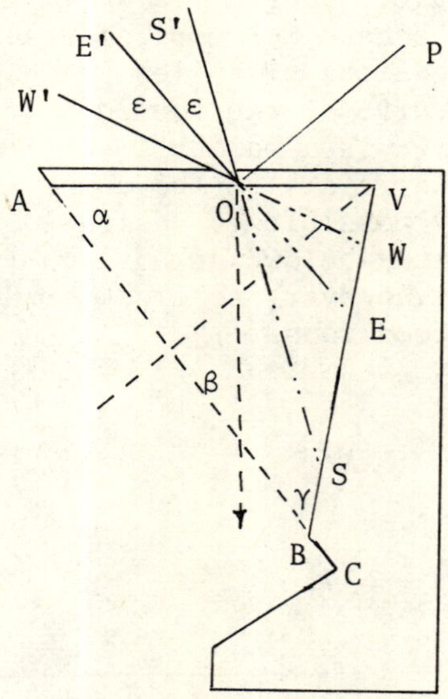

Figure 21

formed by rays of sunlight passing through a gnomon point on the surface of the cone in the meridian plane along the line VA.[16]

The day curves on a roofed conical dial are curves of intersection of two cones with the vertex of one lying on the surface of the other.[17] The equinoctial curve is a circle equally divided by the hour points (see figure 12).[18] From the distance between any two hour points on this circle, say EF, it is possible to find the diameter OE and to use OE and the measured distance VE to find the intended latitude, ϕ, and the angular opening of the cone 2ω. In triangle VOE in figure 21, angle OVE = $2\omega = 2\phi$, angle VEO = angle EOV = $\overline{\phi}$, and

$$\frac{OE}{\sin(2\,\phi)} = \frac{VE}{\sin(90-\phi)} \text{ or } \frac{OE}{VE} = \frac{2\cdot\sin\phi\cdot\cos\phi}{\cos\phi}$$

so that

$$\phi = \sin^{-1}\,[OE/(2\cdot VE)].$$

The distance from the vertex, V, to the gnomon point, O, is equal to VE. On the Izmir dial, both β and γ (see figure 21) can be measured. Their sum is a right angle as it should be if the dial has been correctly identified as a roofed cone.

3. Planar Sundials

Planar sundials of Graeco-Roman origin fall into distinct categories according to the intended position of the flat dial face. They were designed to lie parallel to the horizon, parallel to the prime vertical plane, parallel to the meridian plane, or parallel to vertical planes which lie between the prime vertical and the meridian. On all these various dials, the hours were marked by the shadow of a gnomon point situated outside the plane of the dial face.

a. Horizontal Dials

Fifteen of the forty preserved planar sundials were designed to be horizontal. Figure 22 shows the grid of lines

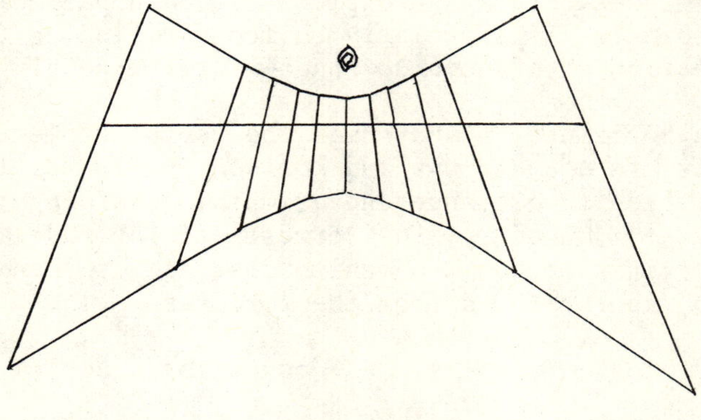

Figure 22

as found on most of these horizontal dials consisting of eleven hour lines, equinox, and the two solstitial curves. The day curves are, of course, plane sections of the sunlight cones. At all latitudes less than 90°, the equinoctial curve is a straight line. At latitudes below 66° (assuming an obliquity of 24°), the remaining day curves are hyperbolic.[19] The eleven hour lines traced by the gnomon point during the year[20] are curves of intersection of the "undulating cone" described by Davies[21] and the dial plane. It is obvious from preserved examples of horizontal dials that straight lines which connect hour points on the summer solstice, equinox, and winter solstice served to approximate these hour lines in Graeco-Roman antiquity.[22]

As figure 22 shows, the solstitial day curves on almost all preserved horizontal dials have been approximated by broken lines which connect the hour points. This seems to indicate that the dialler located these hour points on the dial face before engraving the day curves. We have some evidence that diallers used graphic projection methods to determine the position of the hour points. Both Vitruvius and Ptolemy describe analemmas which for given solar positions serve to determine length and direction of the shadow cast by a gnomon on the face of a planar sundial (see Appendix). Specifically, in his book, *On the Analemma*, Ptolemy gives methods for deriving, both by trigonometric and also by graphic means, three pairs of spherical coordinates for the sun relative to a given place on earth, given solar declination, terrestrial latitude, and hour of the day. Though he does not say so explicitly, each pair of spherical coordinates is singularly suited for finding the length and direction of a gnomon's shadow for a type of plane sundial.

For a horizontal planar dial the useful Ptolemaic coordinates are *descensivus* and *horizon*, or in our terminology, zenith distance and azimuth. According to modern spherical trigonometry, for terrestrial latitude, ϕ, and solar declination, δ, the zenith distance, z_H, and the azimuth, A_H, can be found from the formulas

$$z_H = \cos^{-1} (\sin \phi \cdot \sin \delta + \cos \phi \cdot \cos \delta \cdot \cos \tau) \qquad (3.1)$$

$$A_H = \cos^{-1} \frac{(\sin \delta \cdot \cos \phi - \sin \phi \cdot \cos \delta \cdot \cos \tau)}{(\sin z_H)} \qquad (3.2)$$

where $\tau = n/6 \cdot \cos^{-1} (-\tan \phi \cdot \tan \delta)$ and n is the number of ancient hours before or after noon. The shadow length, q_H, on a horizontal dial, measured from the projection of the gnomon point on the horizontal dial face, where h is the distance between the gnomon point and its projection (i.e., h is the perpendicular height of the gnomon), is related to the zenith distance by the equation

$$q_H = h \cdot \tan z_H. \tag{3.3}$$

The direction of shadow, i_H, measured at the projection of the gnomon point on the horizontal dial face and from the meridian line to the shadow of length q_H, is equal to the azimuthal angle, A_H.

At noon throughout the year, $\tau = 0$. By substituting this value of τ in equations (3.1), (3.2), and (3.3) at the winter solstice when $\delta = -\varepsilon$, we obtain

$$q_H = h \cdot \tan (\phi + \varepsilon) \text{ and } i_H = 180°;$$

at the equinox when $\delta = 0$

$$q_H = h \cdot \tan \phi \text{ and } i_H = 180°;$$

at the summer solstice when $\delta = \varepsilon$

$$q_H = h \cdot \tan (\phi - \varepsilon) \text{ and } i_H = 180°.$$

In each case the direction is the same, indicating that the noon line always falls on the meridian line passing through the projection of the gnomon point. Differences in the lengths of noon shadows at the solstices and equinox, SE and EW in the meridian section in figure 23, are related by the equation

$$\frac{SE}{EW} = \frac{h \cdot \tan \phi - h \cdot \tan (\phi - \varepsilon)}{h \cdot \tan (\phi + \varepsilon) - h \cdot \tan \phi} \tag{3.4}$$

or

$$\tan \phi = \frac{WE - ES}{WE + ES} \cdot \cot \varepsilon. \tag{3.5}$$

The intended latitude of a horizontal dial can thus be found from measurements of distances between noon hour points.

The computed ϕ and the distance WE or ES can be used to calculate the perpendicular height h of the gnomon since

$$h = \frac{WE}{\tan (\phi + \varepsilon) - \tan \phi} = \frac{ES}{\tan \phi - \tan (\phi - \varepsilon)}. \tag{3.6}$$

No gnomons survive on Greek or Roman horizontal sundials, although in most cases the position of the base of the original gnomon is marked by a lead-filled hole. A simple test can be performed to determine whether this hole, G, coincides with the projection of the gnomon point on the dial face, F (see figure 23), i.e., to find out whether the

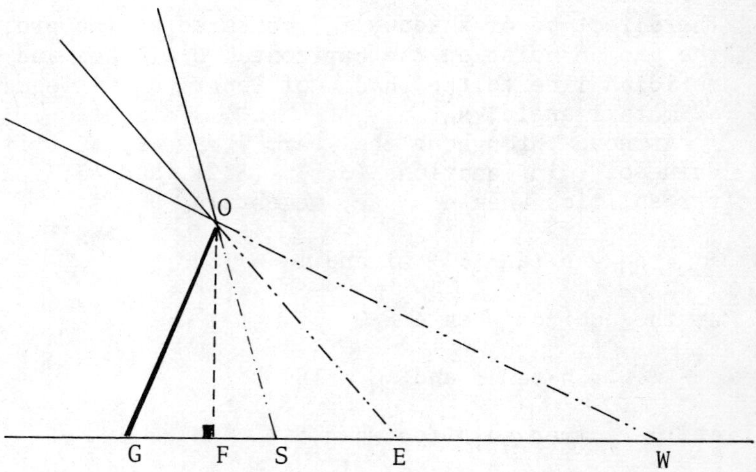

Figure 23

original gnomon was vertical. If the computed length of shortest noon shadow, $h \cdot \tan(\phi - \varepsilon)$, does not equal the measured distance GS on the meridian line, we can conclude that the original gnomon was inclined from the vertical.

It is possible to check the computed values of ϕ and h by using formulas (3.1), (3.2), and (3.3) to find shadow length and direction for any seasonal hour before or after noon and to compare these latter computed values with measurements made on the dial.[23]

b. Prime Vertical Dials

The distinctive network of hour lines and day curves on a horizontal dial is independent of its orientation. However, this same network of lines on a vertical dial varies greatly according to the compass direction which the dial faces. Because the theory of its construction is so close to that of horizontal dials one might expect to find numerous preserved Greek and Roman sundials with surface parallel to the plane of the prime vertical. In fact, as the catalogue testifies, only one carefully constructed south-facing vertical dial is preserved--on the Tower of the Winds in the Roman agora in Athens.

Figure 24 shows the network of lines which appears on the south face of the Tower of Winds. The equinox is a horizon-

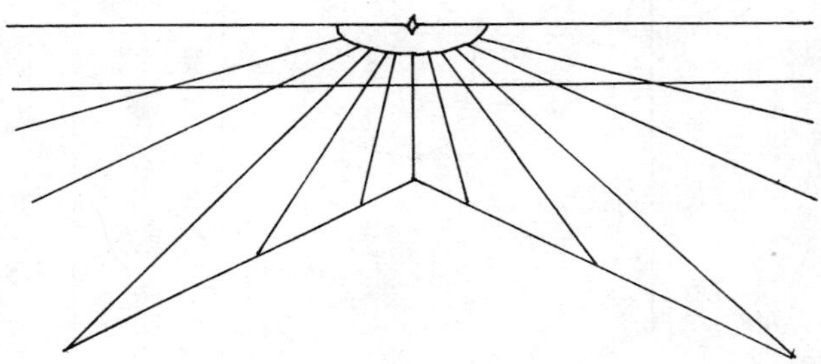

Figure 24

tal straight line as it should be for all latitudes between equator and pole. As plane sections of sunlight cones, the day curves should be parts of hyperbolas for latitudes above the torrid zone.[24] The curve closest to the projection of the gnomon point on the vertical face is that traced by the shadow of the gnomon point on winter solstice. The day curve at winter solstice is limited by a horizontal line passing through the projection of the gnomon point. In spring and summer, the shadow of a gnomon point will fall on a south-facing vertical dial for only part of the day. An additional north-facing vertical dial with its own gnomon point is required for marking some of the seasonal hours between the spring and autumn equinox.[25]

Again it seems likely that the designer of the south (and north) face of the Tower of Winds determined the pattern of hour lines and day curves by constructing seasonal hour points on the dial face and joining them.

The Ptolemaic coordinates relevant to the construction of prime vertical sundials are the *vertical* and *horarius* arcs. Figure 25 shows a celestial sphere with the center at the gnomon point tangent to a vertical dial face. The ver-

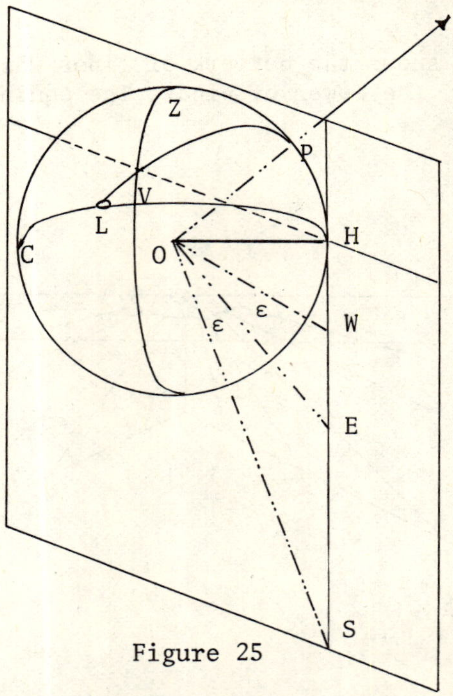

Figure 25

tical arc is ZV and the horarius arc is CL. L marks solar position. Ptolemy's methods for finding the horarius and vertical arcs are equivalent to modern methods which involve the solution of spherical triangle PCL:

$$CL = z_V = \cos^{-1} (\cos \overline{\phi} \cdot \cos \delta \cdot \cos \tau - \sin \overline{\phi} \cdot \sin \delta) \quad (3.7)$$

$$ZV = A_V = \tan^{-1} \frac{\sin \tau}{\tan \delta \cdot \cos \overline{\phi} + \sin \overline{\phi} \cdot \cos \tau} . \quad (3.8)$$

Shadow length, q_V, as defined on the horizontal dial is related to z_V by the equation

$$q_V = h \cdot \tan z_V$$

where h again is the perpendicular length of the gnomon.[26] Direction, i_V, measured at the projection of the gnomon point from the meridian to the shadow of length q_V is equal to Ptolemy's vertical curve, A_V.

At noon, when half daylight, τ, is 0°, the direction angle is 0° at all times of year. The length of noon shadow at winter solstice is $h \cdot \tan (\overline{\phi} - \varepsilon)$; at equinox, $h \cdot \tan \overline{\phi}$; at summer solstice, $h \cdot \tan (\overline{\phi} + \varepsilon)$. As with the horizontal dial, the differences in shadow length may be used to compute intended latitude, since

$$\cot \phi = \frac{SE - WE}{SE + WE} \cdot \cot \varepsilon .$$

Similarly, the perpendicular length of gnomon can be found from differences in the length of noon shadows where

$$h = \frac{WE}{\tan \overline{\phi} - \tan (\overline{\phi} - \varepsilon)} \quad \frac{SE}{\tan (\overline{\phi} + \varepsilon) - \tan \overline{\phi}} .$$

Computed ϕ and h may, of course, be tested by comparing direction and shadow length computed from formulas (3.7) and (3.8) with measured shadow length and direction, making sure to modify the computed values if the base of the gnomon does not coincide with the projection of the gnomon point on the dial face.

Though the only accurate prime vertical sundials from antiquity are on the north and south faces of the Tower of Winds, there are other south-facing vertical sundials from the first centuries of our era which consist of a simple semicircle divided into twelve parts by eleven lines radiating from a central gnomon hole as represented in figure 26. These dials seem to represent a distinct--but more

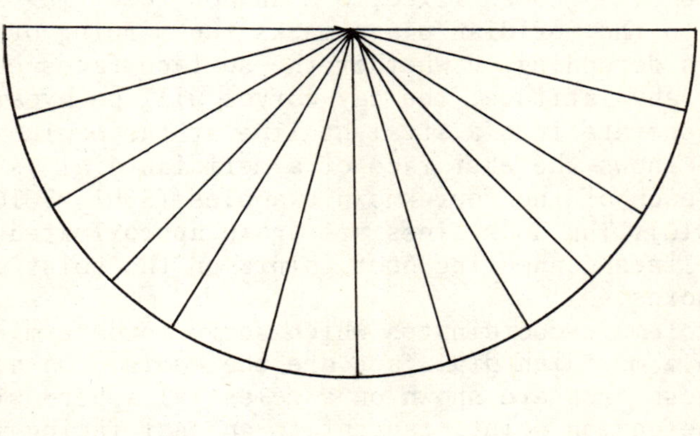

Figure 26

crude--tradition from that which produced the Tower dial, similar to that which produced the Saxon dials of seventh century Britain.[27] The semicircles are unevenly divided. It is difficult to determine what kind of time they were meant to tell. Not a single gnomon is preserved; but the original gnomons probably extended perpendicularly from the vertical dial surfaces. If this was indeed the case, then the radiating lines could have accurately marked the ancient hours on at most two days of the year (i.e. for a single value of δ). However, if the semicircles were divided accurately at hour points on any arbitrary day of the year, the unequal divisions would be arranged symmetrically about the meridian. None of the early examples of the dials seem to exhibit this symmetry. It is quite possible that the unequal divisions are only the result of poor workmanship and that the dial maker intended each of the divisions to represent a twelfth of a circle. It would then be possible to use the divided semicircular grid like a crude protractor to estimate simply the direction of the shadow cast by a horizontal gnomon throughout the day. Late products of the tradition seem to support this suggestion, for they become better protractors rather than better sundials.

c. Meridian Dials

The tip of a gnomon fixed to a shadow-receiving surface parallel to the meridian plane marks the morning or afternoon hours depending on whether the surface faces east or west. For any latitude, the day curves will be hyperbolas which degenerate into a straight line at the equinox.[28] Figure 27 shows the east face of a meridian dial as it appears on each of the four known examples (5001, 5010, 5024, 7001G). The hour lines are again approximated by straight lines connecting hour points on the solstices and the equinoxes.[29]

The Ptolemaic coordinates which serve to determine hour points on a meridian dial face are the *hectemoros* and *meridian*. These arcs are shown on a celestial sphere with center at the gnomon point, tangent to an east facing meridian plane in figure 28. The great circle arc EL is the *hectemoros* arc; the great circle arc BM is the *meridian*. Ptolemy's plane trigonometric determination of these curves is equivalent to the solution of the spherical triangles EPL and LPM where

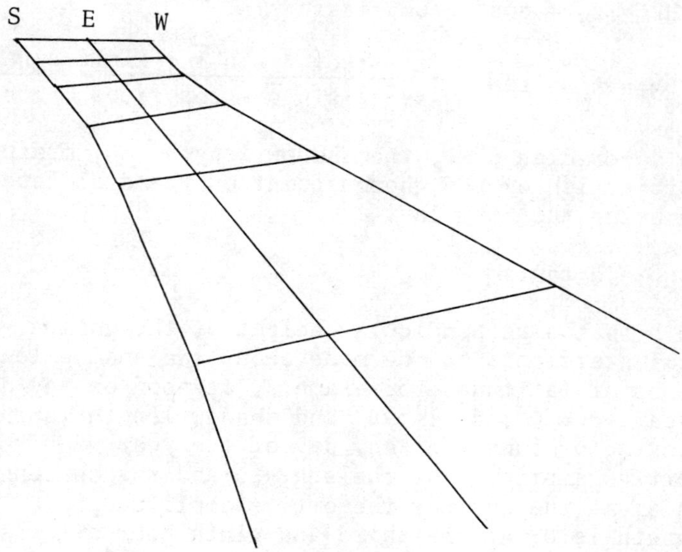

Figure 27

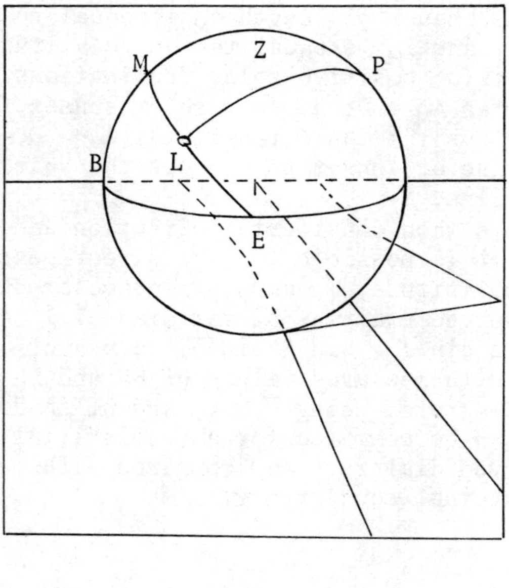

Figure 28

$$EL = z_M = \cos^{-1} (\cos \delta \cdot \sin \tau)$$

$$BM = A_M = \tan^{-1} \frac{\cos \tau \cdot \cos \delta \cdot \sin \overline{\phi} + \sin \delta \cdot \sin \phi}{-\sin \delta \cdot \sin \overline{\phi} + \cos \tau \cdot \cos \delta \cdot \cos \overline{\phi}} .$$

On a meridian dial, the shadow length, q_M, measured from the projection of the gnomon point on the dial face, can be found using the formula

$$q_M = h \cdot \tan z_M$$

where h is the perpendicular height of the gnomon. Certain general statements can be made about the shadow length regardless of latitude. For example, at noon on any day of the year, $\tau = 0°$, z_M is 90° and shadow length is infinite. At sunrise or sunset of any day of the year, z_M is equal to the ortive amplitude of the sun, ξ, and shadow length is $h \cdot \tan \xi$; at the equinox the ortive amplitude is 0° and shadow length is 0. At the third and ninth hour after sunrise on the equinox, δ is 0°, τ is 45°, z_M is 45°, and shadow length is h.

Direction of shadow, i_M, on a meridian dial, measured at the projection of the gnomon point on the dial face from the horizon line to the shadow of length q_M is equal to A_M. At the equinox δ is 0°, and i_M is equal to the colatitude, $\overline{\phi}$. At sunrise or sunset for positive solar declinations, $\cos \tau = -\tan \delta \cdot \tan \phi$, $\tan A_M = 0$; at sunrise or sunset for negative declinations, $\cos \tau = \tan \delta \cdot \tan \phi$, $\tan A_M = 0$. The hour line marking sunrise or sunset throughout the year is a horizontal straight line.

As we have seen above when considering direction and length of the shadow, it is possible to take direct measurements of intended colatitude, $\overline{\phi}$, and perpendicular gnomon length, h, on preserved examples of meridian dials. Since, at the solstices $\sin \varepsilon = \sin \xi \cdot \sin \overline{\phi}$, values of $\overline{\phi}$ and h can be combined with measured values of SE and EW (see figure 27) to estimate ε. Using $\overline{\phi}$, ε, and h, shadow length and direction can be computed for any solstitial hour point on a preserved dial face and compared with measured values to test internal consistency.

d. Deviating Dials

Thirteen Graeco-Roman sundials have vertical planar faces which are not parallel to the prime vertical or the meridian. In some modern discussions of dials, this type of face is said to belong to a declining dial. In order to avoid suggesting an inappropriate association with the celestial coordinate, *declination*, I prefer to adopt a literal translation of Drecker's term *abweichende*, and refer to this group of dials as *deviating*. Except for two examples on the Tower of Winds in Athens, Graeco-Roman deviating dials seem to have been oriented facing southeast or southwest and have been preserved in connected pairs (5008G) or sets of four (5002G).

The pattern of hour lines and day curves on a deviating sundial is shown in figure 29. As is the case with the types

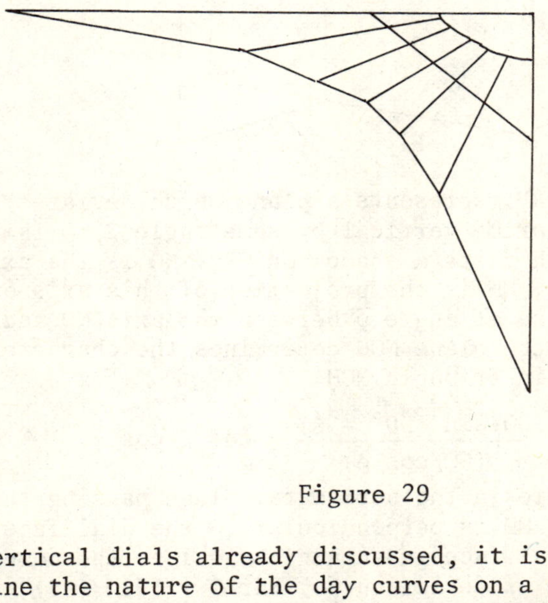

Figure 29

of vertical dials already discussed, it is possible to determine the nature of the day curves on a deviating dial by considering them as conic sections, formed when the receiving surface intersects the sunlight cone. The types of conic sections which result depend upon the orientation of the axis of the cone (parallel to the axis of the earth) with respect to the plane of the intersecting dial face. These determining conditions can be most fruitfully examined in the plane perpendicular to the intersecting plane and passing through the axis of the cone.

In figure 30, XV" represents a plane parallel to the

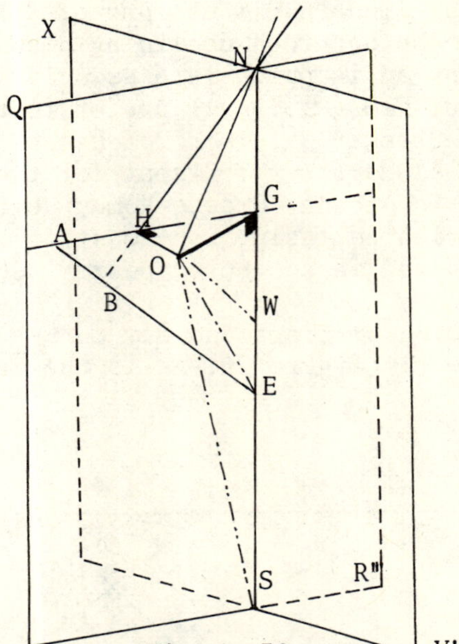

Figure 30

prime vertical; QR" represents a plane which deviates from the plane of the prime vertical by some angle β; O is the gnomon point which casts a shadow on QR". NO is the axis of the sunlight cone; NH is the projection of this axis on the dial face. The size of angle $\overline{\psi}$ between the axis NO and its projection NH in the plane NOH determines the character of the day curves: in triangle NOH

$$\sin \overline{\psi} = \frac{HO}{NO} = \frac{GO \cdot \sin (90^\circ - \beta)}{(GO/\cos \phi)} = \cos \beta \cdot \cos \phi .$$

(Both HO and GO lie in the horizontal plane passing through the gnomon point. HO is perpendicular to the dial face. GO intersects the dial face in an angle equal to the complement of the angle of deviation β.) For $\overline{\psi} > (90^\circ - \delta)$, the day curves will be hyperbolic; for $\overline{\psi} = (90^\circ - \delta)$, the day curves will be parabolic; for $\overline{\psi} < (90^\circ - \delta)$, the day curves will be elliptical. At the equinox, the sunlight cone becomes a plane which intersects the dial face in a straight line. This equinoctial line lies in a plane perpendicular to the projection of the axis of the sunlight cone on the dial face. In figure 30, AE represents the equinoctial line perpendicular to the projection of the axis, NH.

For any vertical dial, including those which deviate from the prime vertical and meridian planes, the noon shadow of the gnomon point always falls on the vertical line of intersection of the dial face with a plane passing through the gnomon. Figure 30 shows this noon line on the deviating dial QR". At the winter solstice, the noon shadow of the gnomon point falls on W; at the equinox, on E; at the summer solstice, on S. From equation (3.4) we have

$$\frac{WE}{ES} = \frac{GO \cdot \tan \overline{\phi} - GO \cdot \tan (\overline{\phi} - \varepsilon)}{GO \cdot \tan (\overline{\phi} + \varepsilon) - GO \cdot \tan \overline{\phi}}$$

or

$$\cot \phi = \frac{SE - WE}{SE + WE} \cdot \cot \varepsilon.$$

Thus the intended latitude of a deviating dial may be found from measurements of differences in shadow lengths at noon. The length of the line GO can be found from formula (3.6), where

$$GO = \frac{WE}{\tan \overline{\phi} + \tan (\overline{\phi} - \varepsilon)} = \frac{ES}{\tan (\overline{\phi} + \varepsilon) - \tan \overline{\phi}}.$$

The distance HO, or the perpendicular height of the gnomon can be found from the equation

$$HO = GO \cdot \cos \beta.$$

For most preserved examples, the only clue to the angle of deviation, β, is provided by the orientation of the equinoctial line. This line (AE in figure 30) meets a horizontal line passing through the projection of the gnomon point on the dial face in an angle (HAB) which is equal to the angle between projected axis and meridian (GNH). In triangle GNH (figure 30)

$$\tan GNH = \frac{HG}{NG} = \frac{GO \cdot \cos (90° - \beta)}{GO \cdot \tan \phi}$$

so

$$\tan HAB = \frac{\sin \beta}{\tan \phi}$$

or

$$\sin \beta = \tan HAB \cdot \tan \phi.$$

Where it is possible to measure the angle HAB (or its complement angle GCB), it is possible to combine the measurement with computed ϕ to find the probable deviation assumed by the dialler.

A good test of the quality of a deviating dial involves computation of the distance along the equinoctial line from noon to the first hour after noon. The distance BE between the noon hour point and the intersection of the projected axis and the equinoctial line can be found by considering triangle BNE in the deviating plane (figure 30) where

$$BE = NB \cdot \tan GNH$$

$$= NB \cdot (\sin \beta / \tan \phi)$$

$$= (NH + HB) \cdot (\sin \beta / \tan \phi).$$

Further

$$BE = [(OH/\tan \overline{\psi}) + (OH \cdot \tan \overline{\psi})] \cdot \sin \beta / \tan \phi$$

$$= \frac{OH \cdot \sin \beta}{\cos \overline{\psi} \cdot \sin \overline{\psi} \cdot \tan \phi} .$$

In triangle BOE in the plane of the equator (see figure 31

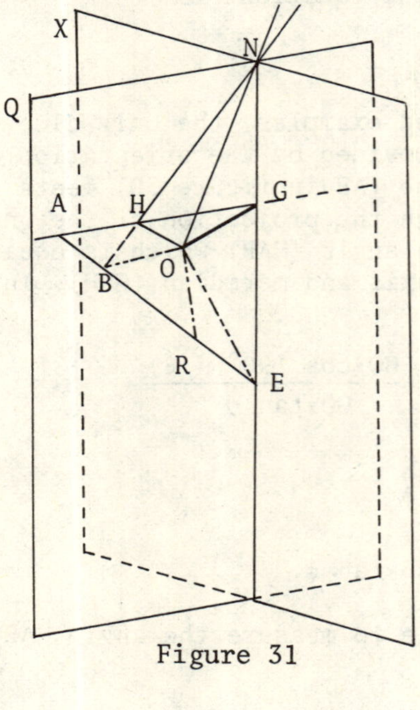

Figure 31

which represents the prime vertical plane intersecting the deviating plane as in figure 30)

$$\tan BOE = BE/BO \text{ where } BO = HO/\cos \bar{\psi}$$

$$= \frac{\sin \beta}{\sin \bar{\psi} \cdot \tan \phi} .$$

If R is the first hour point before (or after) noon, BR can be found by solving the right triangle OBR, where

$$BR = [\tan (BOE - 15°)] \cdot (HO/\cos \bar{\psi})$$

since angle EOR is equal to one equinoctial hour. The computed value RE = BE - BR should nearly be the measured distance RE on the dial face. An analogous test can be made for the distance along the equinox between sunrise (or sunset) and the first (or last) hour point.

While the probable orientation of the shadow-receiving surface of a deviating dial is rather easily determined from preserved evidence, the determination of the orientation and length of its shadow-casting gnomon presents something of a problem. Not a single gnomon is preserved. In some cases the holes which do remain indicate that the original gnomon extended horizontally but not perpendicularly from the dial surface. It is possible that the gnomons which did not extend perpendicularly were always directed toward the south point with their base on the line of intersection of meridian plane and deviating plane. This was certainly the case with those connected pairs of dials which shared one gnomon (5008G and 5002G). One thing which can be calculated with some certainty is the perpendicular length of the gnomon and the position of the projection of the gnomon on the dial face, both from measurements on the noon line--see equations (3.4), (3.5), and (3.6)--and from measurements on the horizon. Figure 32 shows the intersection of a horizontal plane through the gnomon point and the deviating plane of a deviating dial. HO is the perpendicular length of the gnomon; A_S, A_E, A_W mark the shadows of the gnomon point at sunrise, and ξ is the ortive amplitude of the sun at the solstices. Now, the distance $HA_S = HO \cdot \tan (90° - \beta + \xi)$ and $HA_W = HO \cdot \tan (90° - \beta - \xi)$, so that

$$HO = \frac{A_S A_W}{\tan (90° - \beta + \xi) - \tan (90° - \beta - \xi)} .$$

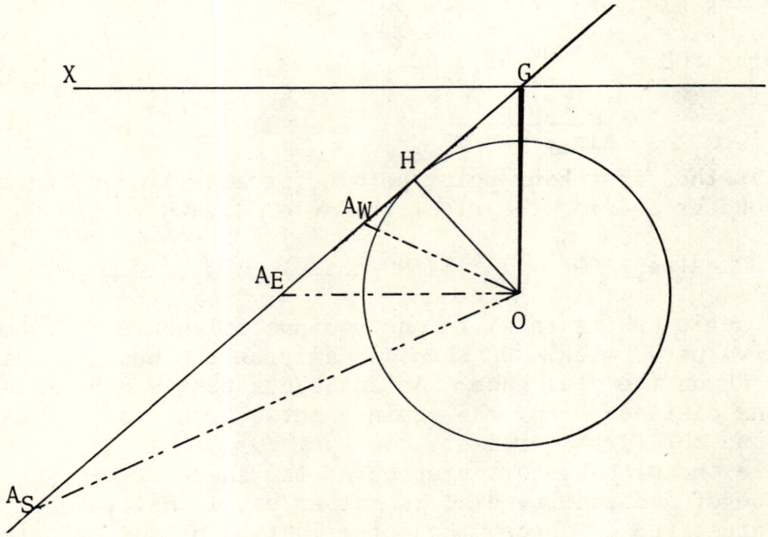

Figure 32

As above, GO is the horizontal line connecting the gnomon point to the noon line. The distance GH from the noon line to the projection of the gnomon point on the dial face can be found by solving right triangle GOH where

$$HG = HO \cdot \tan \beta \cdot$$

The hour points on a deviating sundial cannot easily be found from a pair of Ptolemaic coordinates, since by definition the deviating dial's face is not parallel to one of Ptolemy's coordinate planes. It does, however, seem likely that diallers had some analemma which helped them to project hour points into position on a deviating dial. In his *Histoire de l'Astronomie Ancienne*, Delambre introduces two celestial coordinates, similar to the Ptolemaic coordinates, which would be useful in the construction of deviating dials. The coordinates, great circle arcs FL and ZD, are shown in figure 33, which represents a celestial sphere with center at the gnomon point and tangent to a deviating dial face. The shadow length, q_D, measured from the projection of the gnomon point on a deviating dial is given by the equation

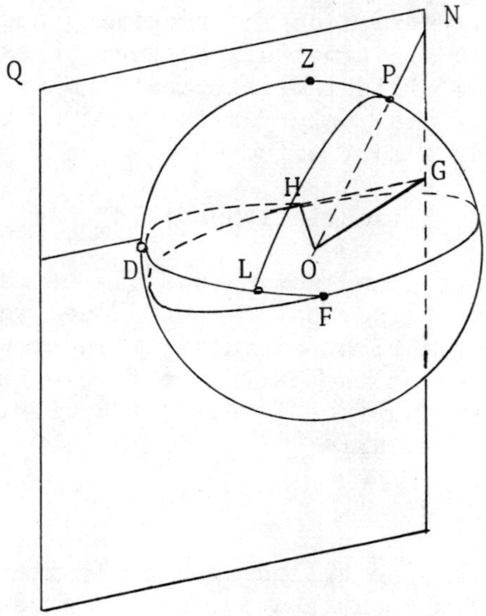

Figure 33

$q_D = h \cdot \tan FL$ (in degrees).

The direction measured at the point G from the noon line to the shadow of length q_D is equal to arc ZD in degrees. It is possible to determine these relevant arcs from three Ptolemaic coordinates, since

$$\cos FL = \cos(90° - \beta) \cdot \cos(\text{hectemoros}) + \sin(90° - \beta) \cdot \sin(\text{hectemoros}) \cdot \cos(\text{meridian})$$

and

$$\cos ZD = \cos(\text{vertical}) \cdot \sin(\text{horarius}) / \sin FL$$

The relationships expressed in the above equations have been found by solving spherical triangles. Ancient diallers could have used the plane or graphic equivalent of this procedure to find the arcs described by Delambre.

Deviating dials share the problem of other vertical dials in that for certain hours of the day at certain times of year the shadow of the gnomon point will not fall on the plane to which the gnomon point has been attached. However, if the angle of deviation β is greater than the greatest ortive amplitude of the sun, the hourly shadow of the gnomon point will always fall on the receiving plane for at least half a day at all times of the year. Preserved examples of deviating dials indicate that the ancients employed this type of dial only for the part of the day during which it would be most effective.

4. Cylindrical Sundials

The corpus of Greek and Roman sundials contains two types of dial with cylindrical face. On one example (6002G) the face is part of a right circular cylinder with axis parallel to the axis of the earth. On five other examples, the shadow receiving surface is part of a right circular cylinder with vertical axis.

a. Inclined Cylinders

The day curves on the cylindrical dial with axis parallel to the axis of the earth are circular arcs parallel to the plane of the equator. The solstice curves are equidistant from the equinox. These characteristics indicate that the gnomon point coincides with a point on the axis of the cylinder. Since a cylinder may be considered a degenerate cone (with $\omega = 0$), most of what has been said about conical dials (with axis parallel to the axis of the earth and gnomon point on the axis) applies here as well. The hour lines actually traced by the gnomon point are again, in practice, approximated by the conic sections which connect hour points equally dividing the day curves.

The radius of the cylindrical surface, r, is the same as the radius of the equinoctial day curve, $OE = 12 \cdot \ell_E/\pi$, where ℓ_E is the arc length between two consecutive equinoctial hour points. The intended latitude of an inclined cylinder can be determined from measurements of arc lengths between consecutive hour points on the equinox and on one other day curve, say the winter solstice, ℓ_W. Half daylight in degrees at winter solstice is θ; half daylight at the equinox is 90°. Since the radius of the winter solstice day curve is equal to the radius of the equinoctial day curve (refer to the meridian section in figure 34),

$$\theta/6 \cdot \ell_W = 90°/6 \cdot \ell_E \text{ or } \theta = 90° \cdot \ell_W/\ell_E.$$

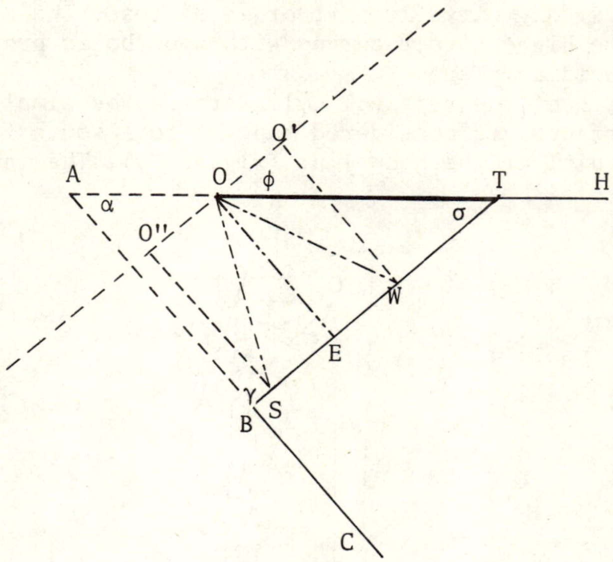

Figure 34

Substituting this value of θ in the trigonometric formula for half daylight at winter solstice, we obtain

$$\cos\,[90°\cdot \ell_W/\ell_E] = \tan\,\phi\cdot\tan\,\varepsilon$$

or

$$\phi = \tan^{-1}[\cos\,(90°\cdot \ell_W/\ell_E)\cdot\cot\,\varepsilon].$$

An analogous formula can be derived using measurements of arc lengths between consecutive hour points on the equinox and the summer solstice.

When the cylindrical surface is cut by planes parallel to the equator and the horizon as in figure 34, then γ should equal 90°, σ should equal ϕ, and α should equal $\overline{\phi}$.

b. Vertical Cylinders

Five Roman sundials have been found with vertical cylindrical faces. The day curves and hour lines on this type of dial are complicated curves which are not simplified by placing the gnomon point in any particular position. The equinoctial day curve is part of an ellipse. The solstitial curves are higher order curves with hyperbolic projections in the meridian plane.

Analysis of the vertical cylinders can be simplified if their surfaces are considered tangent to a south-facing vertical dial at the noon line (figure 35). The intended

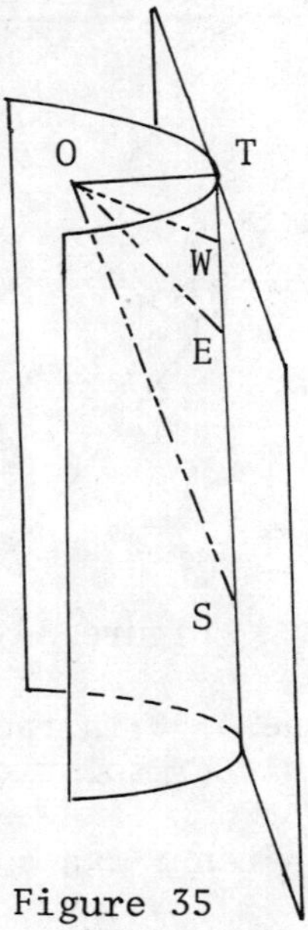

Figure 35

latitude and the distance between the gnomon point and its projection on the noon line can be found from differences in noon shadow lengths as in equations (3.5) and (3.6):

$$\cot \phi = \frac{SE - WE}{SE + WE} \cdot \cot \varepsilon$$

and

$$h = \frac{WE}{\tan \overline{\phi} - \tan (\overline{\phi} - \varepsilon)} = \frac{SE}{\tan (\overline{\phi} + \varepsilon) - \tan \overline{\phi}} .$$

(W, E, and S are the noon hour points on the solstitial and equinoctial day curves.)

3 Vitruvius

Hemicyclium excavatum ex quadrato ad enclimaque succisum Berosus Chaldaeus dicitur invenisse; scaphen sive hemisphaerium dicitur Aristarchus Samius, idem etiam discum in planitia; arachnen Eudoxus astrologus, nonnulli dicunt Apollonium; plinthium sive lacunar, quod etiam in circo Flaminio est positum, Scopinas Syracusius; *pros ta historumena,* Parmenion, *pros pan clima,* Theodosius et Andrias, Patrocles pelecinum, Dionysodorus conum, Apollonius pharetram, aliaque genera et qui supra scripti sunt et alii plures inventa reliquerunt, uti conarachnen, conicum plinthium, antiboreum. Item ex his generibus viatoria pensilia uti fierent, plures scripta reliquerunt. Ex quorum libris, si qui velit, subiectiones invenire poterit, dummodo sciat analemmatos descriptiones.

Vitruvius, *De architectura,* 9.8

Vitruvius's list of sundials and inventors has long provided material for commentary. August Rode, Albert Rehm, E. Ardaillon, Hermann Diels, Joseph Drecker, Jean Soubiran, and, most recently, Derek J. de Solla Price,[1] have all attempted to identify preserved sundials with the types which Vitruvius names. Their success has been limited by the lack of a catalogued corpus of ancient dials. In the process of preparing such a catalogue I have been able to reconsider and perhaps extend the work of these scholars in relating archeological and literary evidence to trace the history of dialling in Graeco-Roman antiquity.

The first dial that Vitruvius mentions in his list of dials and inventors is the *hemicyclium excavatum ex quadrato ad enclimaque succisum*. The Latin phrase seems to refer to a spherical dial with oblique front surface cut according to the latitude of its location like that described in section 2.b of chapter 2 and illustrated in profile in figure

2. One of the first published reports of the excavation of an ancient sundial describes a dial of this type[2] and subsequent archeological finds have shown that many such cut spherical dials were in use in Vitruvius's time (see catalogued dials 1025G, 1020G, 1019G, 1026G from Pompeii and 1032G, 1033G from Ostia). The inventor Berosus Chaldaeus is probably the Babylonian astronomer who is said to have flourished on the Greek island of Cos about 270 B.C. The mention of Berosus in this context is consistent with the statement of Herodotus that the Greeks learned about the πόλος and γνώμων from the Babylonians (see chapter 1).

The *hemisphaerium* or *scaphe* seems to refer to a sundial with a hemispherical surface like that described in section 2.a of chapter 2. Catalogued examples of this design include 1071G, 1070G, 1073, 1072G, 1068G, 1067G. The inventor of this type of dial, Aristarchus of Samos, can be identified with the Greek mathematician and astronomer who is credited by Ptolemy (*Almagest* 3.1) with having observed the summer solstice in 281/280 B.C.[3] A second reference to ἡ τοῦ ὡρολογείον σκάφη occurs in Cleomedes' first century account of the equipment used by Eratosthenes to determine the dimensions of the earth in the third century B.C.[4] The instrument enabled Eratosthenes to measure directly the noon shadow's length as part of a great circle. It certainly must have consisted of some portion of a spherical shadow-receiving surface.

Aristarchus is also mentioned as the inventor of the *discus in planitia*. It is generally agreed that the term refers to a dial with planar shadow-receiving surface; but here the agreement ends.[5] The corpus of ancient dials contains only two examples of meridian dials (see chapter 2, section 3.c) which are not parts of dials with multiple faces. These two similar meridian dials (5024, 5010) are the only known nonportable disk-shaped sundials.[6] It seems likely that they represent the *discus* design of Aristarchus.

Many commentators on Vitruvius's list have suggested that the terms *arachnen* and *conarachnen* refer to the network of hour lines and day curves on a dial face rather than to a particular type of shadow-receiving surface.[7] *Arachnen* derives from the Greek ἀράχνη, meaning "spider's web," and seems an appropriately descriptive term. The suggestion that the terms refer to some sort of metal fretwork[8] has not, to my knowledge, been supported by archeological evidence. The inventor has been identified as Eudoxus of Knidos, the mathematician and astronomer who flourished about 370 B.C.[9] He is the earliest inventor mentioned by Vitruvius, and the singularly high quality of his mathematical works makes it plausible that he made basic contributions also to the theory of dialling.

The *arachnen* is alternatively attributed to Apollonius (probably the mathematician of Perga) who wrote the great treatise on conic sections. Apollonius was active in the late third century and early second century B.C.[10] He is also credited with the invention of a type of sundial called *pharetra*, "quiver." Scholars agree that it refers to a plane deviating dial like those described in chapter 2, section 3.d. Hermann Diels has observed that the elongated nature of the network of lines on a connected pair of deviating dials (see 5008G and 5002G in the catalogue) resembles the flattened out surface of a quiver.[11] G. Rayet justifies the connection by noting that the divergent hour lines seem to be spilling from an upended quiver.[12]

So many roofed spherical sundials date from the first centuries of our era that it seems likely that Vitruvius knew of them. They are distinctive enough to be named as a separate type in his brief list. However, no agreement has been reached on which of the listed names should be applied to this type. It is true that the term *lacunar* can mean "ceiling," and Vitruvius ascribes a dial of that name to the inventor Scopinas of Syracuse. Scopinas's dial is also called *plinthium* (or possibly *panthium*), either of which names may suggest a dial of the roofed type.[13]

The close resemblance between its network of lines and a double-edged axe (πέλεκυς or *bipennis*) has led most experts[14] to apply Vitruvius's term *pelecinum* to a horizontal sundial as described in chapter 2, section 3.a. Certainly no other type of dial in the catalogue calls the double axe to mind. Thus it is possible to identify the *pelecinum* with extant examples. Vitruvius lists Patrocles as the inventor of the *pelecinum*. In a recent commentary on the passage, J. Soubiran suggests that this Patrocles might be the geographer who flourished between the last half of the fourth century and the first quarter of the third century B.C. and whose work was used by Eratosthenes and cited by Strabo and Pliny.[15] Other commentators, perhaps discouraged by the geographer's early date, have declined to associate the two names, and none offer any alternative suggestions.

The *conus* listed by Vitruvius as a type of sundial known in his day can almost certainly be identified with the conical dials described in chapter 2, section 2.a. Vitruvius attributes the invention of the *conus* to one Dionysodorus who has been identified by some scholars as the mathematician of Caunus[16] and by others as the geographer of Melos.[17] Dionysodorus of Caunus, the author of a lost work, Περὶ τῆς σπείρας, *On the torus*, is known for a solution to the problem of cutting a sphere into segments having a given ratio.[18] His solution involved the use of conic sections, so that it seems possible that he would have been capable of solving the problem of designing a conical dial.

Dionysodorus was at work during a time when cones and segments of cones were objects of continuing study. From the writings of a scholastic commentator on mathematical works, Eutocius of Ascalon, we know that plane sections of conical surfaces were first studied in the fourth century B.C. by the geometer, Menaechmus.[19] According to Eutocius, the cones studied by Menaechmus and fellow mathematicians up to Apollonius were right circular cones generated by right triangles revolving about a side adjacent to the right angle. Conic sections studied during this period were those formed by a cutting plane perpendicular to the hypotenuse of the triangular generator of the cone. The resulting conic section might be a parabola, a hyperbola, or an ellipse depending on whether the vertex of the generated cone was a right angle, an obtuse angle, or an acute angle. There is no evidence in the literature that this definition of the conics changed before Apollonius published his work about 200 B.C. Apollonius relaxed the terms of the early definition in two ways: he considered *any* circular conical surfaces, even skew ones, cut by *any* plane.[20]

As E. J. Dijksterhuis has pointed out, the celebrated mathematician, Archimedes (died 212 B.C.) still considered conic sections as they had been defined by Menaechmus,[21] but was well aware of the more general methods of generating cones later published by Apollonius. The fact that right circular conical sundials with faces arbitrarily sectioned by planes parallel to the celestial horizon and the equator are known to have been made before 200 B.C. further suggests that Apollonius's work is the culmination of developing ideas concerning conic sections. It is, in fact, tempting to suggest that developments in dialling influenced developments in geometry in this period.

The date of Dionysodorus of Caunus (end of the third century B.C.) is very close to the date of the earliest known conical sundial, 3049G in the catalogue. The dial was discovered in 1873 by a French expedition to Heraclea ad Latmum near Miletus in Turkey. An inscription on the dial states that it was made by Themistagoras, an Alexandrian dialler. It is dedicated to King Ptolemy. In analyzing the inscription, G. Rayet remarks that there are two possible Ptolemies to whom the dedication might refer.[22] The first, Ptolemy Philadelphus (285-246 B.C.) is known to have been a patron of the sciences; the second, Ptolemy Euergetes I (246-221 B.C.) is known to have conducted military campaigns in Asia Minor. The workmanship of the dial is very fine, suggesting a relatively early origin. On the other hand, the two possible identifications of the inventor Dionysodorus suggest that the dial probably dates from a period later than the reign of either Ptolemy.

We have no evidence that Vitruvius was familiar with the type of roofed conical dial described in chapter 2, section 2.c. It is possible, however, that he does name it in his list of various dials, for he mentions a *conicum plinthium*, without specifying its inventor. If, as I have suggested, the term *plinthium* refers to a roofed spherical dial, then the *conicum plinthium* seems naturally to refer to a roofed cone. In fact Granger translates the phrase "conical ceiling" in the Loeb Classical Library edition of *De architectura*.

Most scholars agree that the shadow-receiving surface of the *antiboreum* dial was designed to face north.[23] They note that no example stands alone, but each is in combination with a dial which faces south. Professor Ionescu-Carligel[24] suggests an interesting variant interpretation of the term which he takes to mean any conical dial with vertex below the horizon (i.e. opposite the north pole) whether it faces north or south (see chapter 2, section 2.b). The fact that the *antiboreum* dial is listed with other conical variants lends credibility to Carligel's statement. However, the corpus contains a north-facing spherical dial to which the name *antiboreum* can also be properly applied.

Three types of sundials mentioned by Vitruvius have traditionally been considered portable. Derek J. de Solla Price lists and classifies eleven known examples.[25] He suggests that the *pros ta historumena* invented by the architect Parmenius might be identified with a type of dial consisting of a set of plates like an astrolabe, the best example being one now preserved in the Kunsthistorisches Museum in Vienna. The term *pros pan clima* suggests a type of dial which allows for a continuous variation of latitude. Four similar examples of this type of dial are known. The best preserved is that now in the Museum of the History of Science, Oxford. The inventor, Theodosius, has been identified as the author of extant works on spherical trigonometry. Andrias is otherwise unknown. The *viatoria pensilia* mentioned by Vitruvius seems to refer to hanging dials in general. Price applies the term to two portable dials which cannot be assigned to the other two groups.

If one considers the dates of the inventors listed by Vitruvius as an indication of the developmental sequence of types of dials, one arrives at something like simultaneous invention of horizontal, meridian, and hemispherical dials, followed by the invention of the cut spherical dial and later by the invention of deviating and conical sundials. Archeological evidence suggests that design of the roofed spherical dial, with the earliest known example dating from the first century B.C., was probably one of the latest contributions to Graeco-Roman dialling. The catalogued evidence also shows that there was no obvious development or modification of a type of dial once it had been invented and a construction procedure had been established (see chapter 4); if anything, dial types show signs of degeneration with time.

Vitruvius includes in *De architectura* 1.6 a description of a very simple procedure for locating the meridian which could have been applied to the problem of properly orienting a sundial. The procedure involves marking the shadow of a vertical gnomon on a level horizontal surface. The shadow of the gnomon point is marked sometime before noon and, with the base of the gnomon as center, a circle is drawn passing through this point. The point where the afternoon shadow touches the circle is marked. The arc between these two shadow points is bisected and the bisection point is connected to the base of the gnomon to form the meridian line. Although Vitruvius does not mention the possibility, it is true that a sundial to be oriented could then be placed so that its noon line was in the vertical plane passing through this meridian line. Facing north or south according to its design, the dial would then be properly placed.

In addition to fixed and portable sundials, Vitruvius mentions two other timekeepers known in his day. These are the *horologium ex aqua*, or water clock, and the *horologium anaphoricum*, or anaphoric clock. Both marked the seasonal hours.[26] Evidence for a noninstrumental method of marking seasonal hours exists in medieval and Byzantine shadow tables that show arithmetical patterns which, as O. Neugebauer has recently pointed out,[27] probably originated in Greek times and certainly were known in the Ptolemaic period in Egypt. The numerous and various preserved stone sundials, however, comprise the most thoroughly documented method of marking the seasonal hours in Graeco-Roman antiquity.

4 A Comparative Summary

This comparative study is a summary of the accompanying catalogue of Greek and Roman sundials. For each type of dial face--spherical, conical, planar, and cylindrical--known provenances and datable examples are listed; materials, dimensions, accuracy, and internal consistency are discussed; possible construction procedures are described; and unusual variants are noted. Inscriptions and decorative details are discussed in a special section. Finally, I have included some comments about the two most important sources of ancient dials, Pompeii and Delos.

Spherical Sundials--1001-1075 and 2001-2023

Greek and Roman diallers produced a surprisingly heterogeneous group of sundials with spherical faces of which 98 survive. They adopted various methods of marking the shadow of a gnomon point which was not always placed at the center of the sphere. The greatest number of sundials with spherical dial faces have been found at Roman sites in Italy (Aquileia, Città Lavinia, Larino, Ostia, Palestrina, Pompeii, Riano, Rome, Terracina, Tivoli, Velletri), in Spain (Merida), in Africa (Carthage, Cherchel, Lambèse, Leptis Magna, Volubilis), in Sicily (Castroreale, Catania), and in parts of Europe not on the Mediterranean (Cannstadt, Grenoble, Nissan). (See figure 36.) Spherical dials have also been found in Asia Minor at Pergamon and Ephesus; in Syria at Cippe de Dmeir; in Egypt at Naukratis; in Dacia at Istropolis, Tomi, and Dorostorum; in Thrace at Istanbul, Thessalonica, and Loutra; in Dalmatia at Metulum; in Greece at Kea, Samothrace, Delos, and Olympia. The dials were found in the ruins of private houses in Pompeii (1020G), Herculaneum (1031G), Nîmes (1045), and Nissan (1046). A fragment from Delos was found in the agora (1002). The spherical dial from Merida (1043) was found in a theater, and a spherical dial from Ostia (1032G) was found at the baths of Neptune.

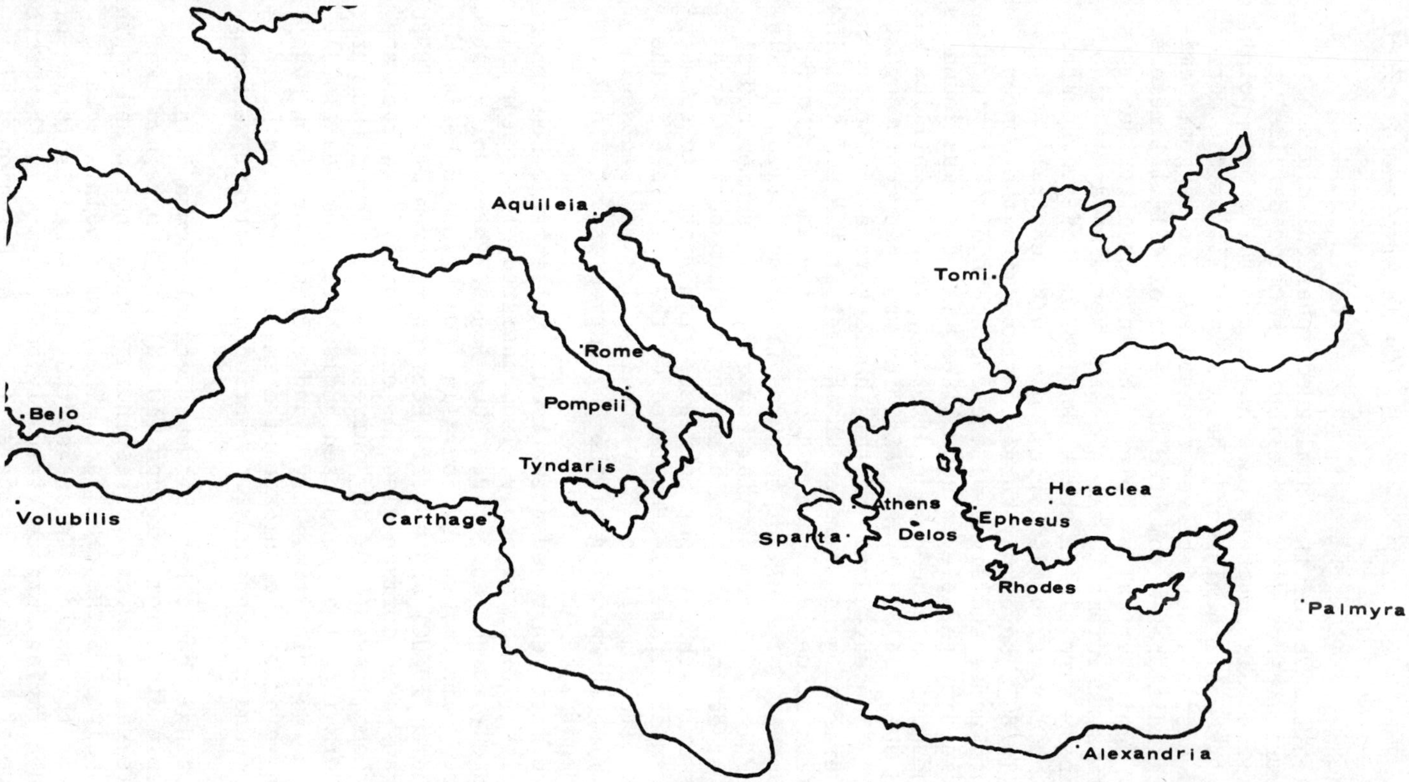
Aquileia
Tomi
Rome
Belo
Pompeii
Tyndaris
Heraclea
Athens
Volubilis
Carthage
Ephesus
Sparta
Delos
Rhodes
Palmyra
Alexandria

Figure 36

At least seven of the catalogued spherical dials have a hemispherical face with top circular section parallel to the horizon. Six of these once had a central gnomon point. Two of these six hemispheres have been split at the meridian, and east and west halves transposed. The largest hemispherical dial has a computed diameter of 451 millimeters. For all hemispherical examples, the intended latitude, computed from measurements taken on the meridian and equinoctial lines, agrees with the known latitude of provenance.

On the largest hemispherical dial with central gnomon point (1068G), seven day curves, marked with the appropriate month names and signs of the zodiac in Greek letters, are divided by the eleven hour lines. The dial was found in Rome and must have been made after the month Quintilius was renamed July (44 B.C.) and Sextilius was changed to August (9 B.C.). The day curves on this hemisphere are also crossed by two arcs of great circles which meet at the point of intersection of meridian and winter solstice and mark off equal amounts of daylight measured from the horizon (6 winter hours = 4 2/5 equinoctial hours = 3 1/2 summer hours). The intersecting arcs form a triangle which contains the amounts by which daylight increases from winter to summer.[1] The circular figure also engraved on the dial face is the *menaeus* circle mentioned by Vitruvius. Its usefulness in simplifying the placing of the day curves is explained in the Appendix.

The damaged engraved network on a Hellenistic hemisphere from Delos (1071G) probably also consisted of seven uninscribed day curves and eleven hour lines. Three uninscribed day curves are divided by hour lines on a second Delian hemisphere (1070G) and a third hemisphere from Delos (1072G) with transposed surface also has only three day curves and eleven hour lines. Four day curves are visible on the published drawing of a transposed hemisphere from Kea (1073). Only the two solstitial curves and eleven hour lines have been engraved on the surface of a small fragmented hemisphere found at Cannstadt, a section of Stuttgart, Germany (1067G).

The single hemisphere without central gnomon point is the double dial from Pergamon (1069G) which once had two separate gnomon points corresponding to the north and south facing halves of the dial face. The gnomon points were placed in the meridian plane so that the summer solstice of the south-facing surface virtually coincided with the winter solstice of the north-facing surface. The Pergamon dial is the only completely preserved untransposed hemisphere.

Most spherical sundials are well characterized by Vitruvius's phrase *hemicyclium excavatum ex quadrato ad enclimaque succisum*. Their faces are bounded by a plane parallel to the horizon and a plane parallel to the celestial equator. The angle between these limiting planes is a measure of the colatitude of the intended location of the finished dial. The spherical segment between the planes is usually engraved with three day curves and eleven hour lines. Two preserved examples have more than three day curves (1001, 1004); four examples have only two day curves (1011, 1040G, 1041, 1042); six examples have no day curves at all (1005G, 1006G, 1021G, 1047, 1052, 1063G).

A particularly interesting fragment of a cut spherical dial has been found at Istropolis (1044). Professor Ionescu-Carligel, who has seen and measured the fragment, has determined that the portions of equinox and summer solstice curves visible on the dial surface have been accurately placed according to the latitude of its known provenance. At least thirteen additional markings on the preserved meridian line indicate solar position and have considerable calendrical significance. An inscribed portion of "daylight triangle" is also visible on the dial fragment. E. Popescu reports that the workmanship and epigraphy of the fragment can be attributed to the third century B.C. If this is so, it is the earliest known spherical sundial.

The positions of the seasonal markings on the early Istropolis fragment are remarkably exact. Later examples of cut spherical dials (see chapter 2, section 1.b) are divided by "day curves" which only look like solstices and equinoxes. These curves are not solstitial or equinoctial curves but arcs of arbitrary circles parallel to the equator. Hour lines should divide these day curves into equal arcs. Sufficiently accurate hour lines can also be constructed by dividing the oblique front edge of a properly cut sphere, into twelve equal arcs and connecting division points to the foot of the gnomon with great circle arcs. It seems likely that the parallel circles on late spherical dials served primarily to locate the hour lines and only secondarily (if at all) to estimate the season of the year.

A small group of spherical sundials consists of examples with quarter spherical dial faces having top edge in a plane parallel to the horizon and front edge in a plane parallel to the plane of the prime vertical circle. Some representatives of this type of dial have three engraved day curves which lie in planes parallel to the front edge (1048G and 1049). On two examples, the eleven hour lines divide only one day curve (1050G, 1051G). One example from Carthage (1052) has no day curves; its eleven hour lines divide the front edge. The gnomon point is preserved on the Carthage dial; it lies in the center of the spherical surface.

The Carthage quarter sphere functions independently of the latitude. No matter where in the *oikoumene* such a dial is placed, it tells the seasonal hours poorly. Dials with similar dubious adaptability have been found near Ephesus (1055G), Thessalonica (1053G), and Dorostorum (1054G). These dials are characterized by a quarter spherical or near quarter spherical dial face divided by eleven hour lines which radiate from the gnomon hole to an engraved circle. The circle is not necessarily related to a particular daily path of the sun. Such dials approximate the seasonal hours in a manner similar to the divided semicircles described in chapter 2, section 3.b.

Cut spherical dials appear to be the products of a very simple construction procedure (figure 37). The dialler

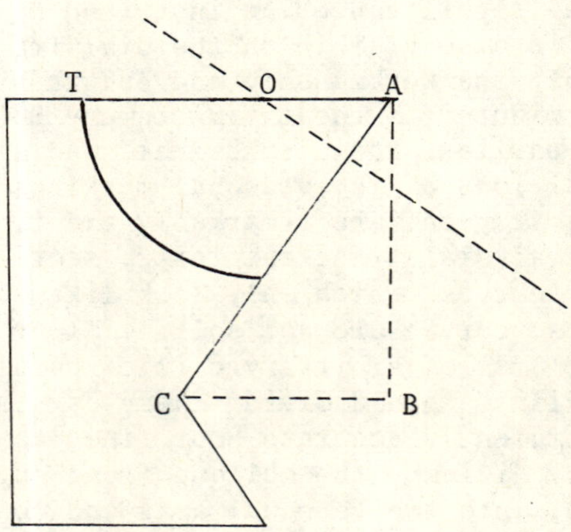

Figure 37

probably began with the value of the angle between horizon and equator for his location. Once the block of stone had been cut according to this angle and the radius of the spherical face had been marked on its horizontal top surface, the dialler probably prepared the spherical hollow. On preserved spherical dials, the angle TAC (figure 37) generally agrees with the complement of the known latitude of provenance. (A spherical sundial at an early stage of construction was found in the Italian agora on Delos [1064] cut according to the local latitude and roughly hollowed.) Once the spherical hollow had been prepared, three arbitrary day curves could easily be engraved parallel to the front opening, divided into twelve equal parts and connected by eleven hour lines. The gnomon of the cut sphere extended horizontally from the center of the top edge so that its point coincided with the center of the spherical surface. The shape of the gnomon hole on many cut spherical dials indicates that the lost gnomons were similar in shape to the elongated pyramids which remain on some conical dials.[2]

So many roofed spherical sundials have been discovered at Aquileia in northern Italy (2002G to 2011G) that the site must have been a manufacturing center. Examples found at nearby cities of Altino (2001G), Trieste (2014G, 2015), Pula (2012G, 2013G), and Salona (2016G) were almost certainly products of Aquileian diallers which were probably exported during the first century A.D. when Aquileia was known for its prosperity. The roofed spherical sundials found at Bologna (2018G), Volterra (2017G), Rome (2020 to 2023G), and Pompeii (2019G) were made by diallers influenced by the Aquileian style. As an important early Roman trading center Aquileia also had connections with the eastern Mediterranean area. A very fine example of a roofed spherical dial has been found on the island of Tenos (7001G) with an inscription identifying it as the work of Andronicus Cyrrestes, the architect who designed the Tower of Winds in Athens in the first century B.C. No other examples of this type of dial are known in Greece or Asia Minor. It may be that one of the ships returning to Aquileia carried an architect or dialler (possibly the Scopinas mentioned by Vitruvius) who was familiar enough with the type of dial uniquely known from Tenos to initiate production of it for other latitudes.

On all preserved examples of roofed spherical sundials, eleven hour lines divide winter solstice, equinox, and summer solstice curves. On the Tenos dial, two additional seasonal indicators divide the meridian between solstices. Similar seasonal markings are visible on other Greek spherical sundials (1001, 1073) but none appear on other roofed spherical examples.

Analysis of measurements on roofed spherical sundials shows that the intended latitudes generally agree with the latitudes of known provenance, although these measurements do not strictly conform to the requirements of theory. Various projection methods must have been used to construct the dials. Diallers who began with a knowledge of the gnomon-shadow ratio of a roofed dial's proposed location could easily set up an analemma corresponding to the dial's meridian section. The summer sunlight cone does not intersect the spherical face beyond the plane which passes perpendicularly through the line which connects the center of the spherical surface and the noon point at the summer solstice (SCT in figure 38). I believe the oblique front plane of

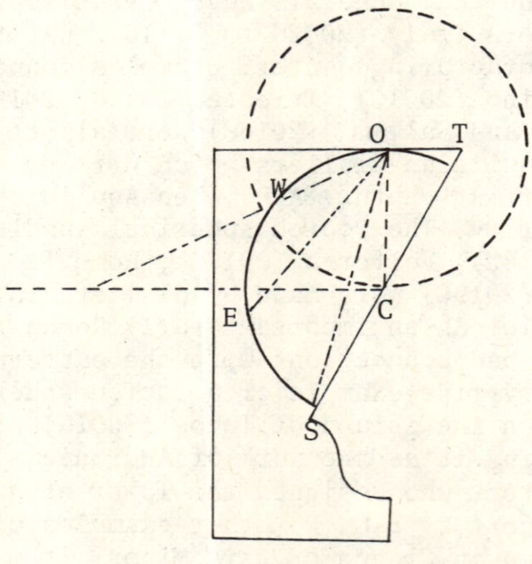

Figure 38

each roofed sphere coincides with this plane, which need *not* be parallel to the equator. Once the position of the oblique front edge had been determined, the stone could be cut and hollowed. The relative positions of the engraved day curves at the meridian could easily be found from the analemma. The experienced dialler would have been familiar with the characteristic pattern of these day curves. He must also have devised some method of dividing the engraved day curves to mark the seasonal hours. I am not certain that he intended divisions on the equinoctial curve to be equal, because this theoretical requirement is not satisfied in many cases. A pattern of alternating distances between hour points on the winter solstice curve is visible on some dials (2003G, 2010G). Distances between hour points on the summer solstice generally decrease from the meridian (see 2005G and 2019G). The pattern created by the hour lines connecting these hour points is always symmetric about the meridian. On roofed dial 2023G, probably from Rome, the third, sixth, and ninth hour lines are marked by a short crosspiece.

As was explained in the section on the theory of roofed spherical dials, the "gnomon point" lies *on* the spherical surface. The hole is not completely preserved on any one example, but traces on three dials (2017G, 2023G, and 2020) suggest the insertion of a pierced metal plate at the highest point of the sphere. The roofs of these three dials have a flat outer surface. On several other examples the outer surface of the partially preserved roof is slightly rounded.

Conical Sundials--3001-3109

One hundred nine sundials preserved from Greek and Roman antiquity have a conical face. They form the largest group in the corpus. Most have been found at sites in the eastern Mediterranean area (see figure 36): Aphrodisias, Athens, Chios, Knidos, Corinth, Delos, Ephesus, Heraclea ad Latmum, Imbros, Izmir, Jerusalem, Kos, Kephalos, Megara, Melos, Memphis, Olympia, Oropus-Amphiareion, Palmyra, Paphos, Pergamon, Piraeus, Rhodes, Samos, Side, Sparta. About twenty-five conical dials have been found in northwest Africa, Italy, and Sicily: at Altino, Carthage, Montagnana, Ostia, Otricoli, Pompeii, Preneste, Rabat, Rome, Tyndaris, and Velletri. A very few examples have been found in Roman Gaul at Aix, Auriol, Auxterre, Avenches, and St. Bertrand de Comminges. Five conical dials were found in private homes. Fourteen were found near temples, theaters, baths, and public marketplaces; four of these latter are monumental in size.

We have noted that the earliest known conical dial, from Heraclea ad Latmum (3049G), may belong to the third century B.C. Inscriptions on dials from Imbros (3044G), Samos (3047), Claros (3051G), and Phoenicia (3094G) date from the second century B.C. All the conical dials found at Pompeii were, of course, made before 79 A.D. A conical dial found near the Roman corinthian temple at Knidos probably belongs to the second century A.D. The remaining conical dials cannot be precisely dated although they bear clear indications that they derive from the Hellenistic period, i.e., from the period between 300 B.C. and 300 A.D.

The three largest known sundials--3008G, 3051G, 3058G--have conical dial faces about three feet high and four feet wide. Most conical dials are one third this size, but a smaller model, six inches high, has been found in the agora in Athens (3010G). White marble is the most common material. All nine complete or partially preserved gnomons are of bronze, but iron stains are visible in at least two gnomon holes. Further evidence indicates that lead was used to fix the gnomons in their sockets.

Eleven hour lines divide each conical dial face. On two conical dials, two extra hour lines mark the beginning of the first hour of the day and the end of the last hour of the day. These extra lines are engraved near the top left and right edges, which edges suffice to mark these times on all other conical dials. On most conical dials the hour lines divide three day curves: winter solstice, equinox, and summer solstice. Additional day curves are divided on 3048 and 3049G. On one dial, 3059G, only the two solstice curves are divided by the hour lines. A single day curve is divided on two conical dial faces, 3004G and 3075. On at least nine dials (for example 3018G) the hour lines divide no engraved day curves at all but radiate from the gnomon hole to the circular front edge of the conical surface.

As explained in chapter 2, it is possible to determine the intended latitude of a conical dial by measuring the angle α, between the horizontal top surface and the oblique front surface. On almost every catalogued conical dial, α is close to the value of the colatitude of the dial's known provenance.On some dials it was possible to measure the angle β, the compliment of α, and hence get a direct measure of the intended latitude.

The intended latitude may also be computed from measurements taken along the meridian line of a conical dial. Latitudes computed in this manner agree to within two degrees of the latitude of provenance for twenty-seven of the catalogued conical dials. For another six, the agreement is within three degrees. For fifteen conical dials on which meridian measurements could be made, the computed latitude differs greatly from the known latitude of provenance and from the latitude measured by the angle β. On dial 3060G from Turkey, for example, the latitude computed from measurements taken along the meridian line is nine degrees greater than the latitude measured by the angle β. The three circular arcs divided by the hour lines are clearly marked χειμερινός, ἰσημερινός, and θερινός, although it seems that they do not accurately mark these specific day curves. Such discrepancies can be explained as the result of carelessness or ignorance. Linear measurements taken along the day curves of catalogued conical dials rarely give an accurate indication of the intended latitude, although the theory requires that they do so. For many conical dials on which day curve measurements could be made, the computed values of ϕ_W and ϕ_S are not equal to each other or to the intended latitude as determined by other methods.

The inconsistencies in some conical dials and the evidence provided by preserved guidelines suggest a possible construction procedure. Before cutting into his block of stone, the dialler probably prepared a simple analemma for the intended latitude based on the gnomon-shadow ratio for that latitude. Figure 39 gives such an analemma for Athens in which angle IAC should be the angle, α, between the horizontal top and the oblique front of the finished dial. AP, AC, and AV are rays of sunlight at noon on the winter solstice, equinox, and summer solstice respectively. Any line intercepting these rays (say QR) may serve as a generator of a conical shadow-receiving surface. Evidence in the catalogue indicates that some diallers preferred to orient QR so that TW = WE. The gnomon length of the finished dial should then be TA; SF, parallel to AC, represents the oblique front surface; ITF, the horizontal top surface. Measurements on the analemma need only be scaled to conform to the desired dimensions of the dial, or to the size of the block of stone to be used.

Preserved construction guidelines clearly visible on 3025G and possibly also on 3023G show that the distances IT, TM, MA were marked by engraving parallel lines on the horizontal top surface of the stone, and this was very likely the first step in applying the analemma. The dialler then cut away the front of the stone along FGS so that angle IFS = angle IAC. He could easily trace the circular base of the cone with radius OS, on the resulting plane surface (a trace of this circular arc is preserved on 3028G). After marking line T and the circular base, the dialler could shape the conical surface, possibly using a device like that described by Rayet in the conclusion of his article.[3] Once the conical surface had been shaped, the meridian line could easily be engraved and the noon shadows

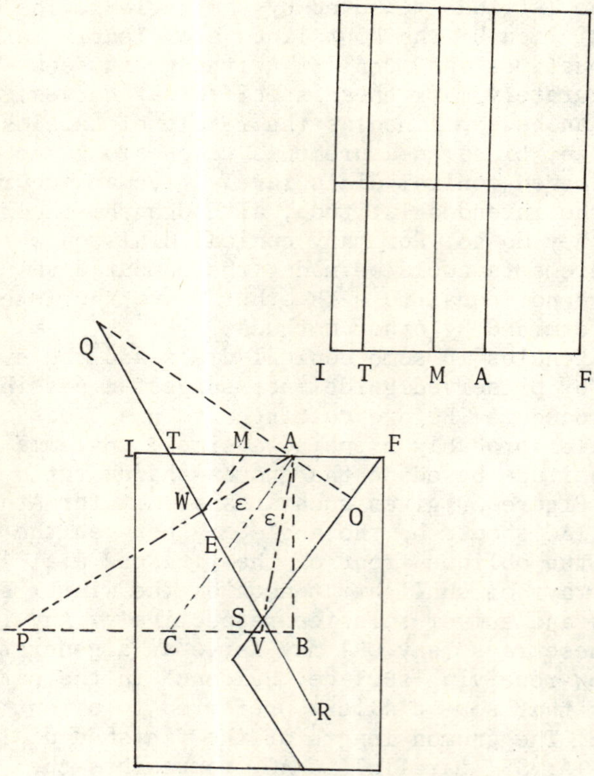

Figure 39

(W, E, S) marked on it. The engraved day curves intersect these shadow points and should follow the curve of the lower edge to meet lines M, A, and F at the upper edge. The dialler undoubtedly placed the ten other hour lines by first dividing the day curves (see the division "dots" on dial 3086G) and then connecting corresponding division points with as straight a line as possible on the conical surface.

By using the analemma, the dialler would have known the required height and depth of stone before beginning. The required width of the finished dial could not be easily predicted using the two-dimensional analemma. In fact it should be slightly larger than twice the radius OS. A mistake in estimating width or in placing the point O would affect the accuracy of day curve divisions and would result in a discrepancy between the value of ϕ computed with these divisions and the latitude intended by the dialler. On many of the dials I examined, I noted that first and last hour divisions were smaller or larger than the other ten equal divisions on a day curve, a circumstance which could reflect a misjudgment of width. On a dial constructed with the aid of an analemma, the best measure of the latitude intended by the dial maker is the angle between the top and front surface; the next best indication of the intended latitude is provided by the relative distances on the meridian line; measurements taken on day curves give the least reliable indication of the intended latitude. After comparing computed and graphically determined parameters of the horizontal conic sections of dials in the Naples museum, Rayet concludes that Greek diallers did not "calculate the dimensions of a conic surface with the idea of obtaining a predetermined horizontal section" since the horizontal sections of the Naples dials always have axes of length greater than that demanded by the other dimensions of the dial.

No working conical dial is complete without a gnomon whose point is on the axis of the shadow-receiving surface. There is evidence to suggest that on all examples the gnomon was fixed horizontally in the meridian plane with its base in the dial's top surface. Preserved gnomons are elongated pyramids (see 3001G, 3023G, 3024G, 3064G). The shapes of gnomon holes indicate that this was the accepted style.

A number of similar conical dials made of the same material (see 3012G) have been found in the Athenian agora, and this may mean that a manufacturing center was located there. A single center of production in the eastern Mediterranean may have been responsible for the distinctively shaped conical dials with rounded sides and a small plinth base, but its location cannot be determined. Examples have been found at Ephesus (3058G) and on the islands of Rhodes (3034G to 3037G) and Kos (3038G to 3040G). With one or two unrelated exceptions, dials found elsewhere have straight back and sides.

Renate Tölle[4] reports the existence of a unique conical dial from Samos, 3046 in the catalogue, with an engraved "daylight triangle" like that on the horizontal dial from Delos, 4001G, and the hemispherical dial now in the Vatican Museum (1068G). On the Samos example, the hours outside the triangle represent equal periods of daylight during the year and display the relationship between seasonal hours.

Two dials included in the catalogue can be characterized as dished cones (1056G, 1060G). The top edge of their dial face is a conic section; the oblique front section is a circular arc; but the meridian line is slightly curved. The difference between chord length WS and arc length WS is so small that I believe the dialler intended to produce a conical dial, but made an error in the first hour line/day curve network, and in erasing his mistake was forced to make the surface concave.

The corpus of Graeco-Roman sundials includes few variant conical dials. Most examples are conventional, south-facing with central meridian line. However, two dials have a conical surface which has been split at the meridian and *transposed*, 3106G from Rhodes and 3107G from Sparta. The Spartan example also has a north-facing transposed conical surface. Untransposed north-facing conical surfaces are part of dials 3049G from Heraclea ad Latmum and 7003G from Tyndaris in Sicily. The most interesting conical variant is in the Basmahane Archeological Museum in Izmir (3109G). Vertex and upper portion of the cone are preserved in the stone, so that the dial has the meridian section illustrated in figure 21. In chapter 3 I identified this example as the *conicum plinthium* of Vitruvius. The day curves intersect the conical surface in a pattern similar to that which appears on roofed spherical dials.

Planar Sundials--4001-4015 and 5001-5025

The forty planar sundials (fifteen horizontal, twenty-five vertical) in this catalogue were found at geographically dispersed sites from Carthage and Borcovicium to Samaria-Sebaste (see figure 36). Most examples were found in Roman Italy and Gaul (Altino, Aquileia, Awans-Biersette, Brugg, Evelette, Narbonne, Ostia, Palestrina, Pompeii, Rome, Syracuse, Tivoli, Wiesbaden). The relatively fewer Greek sites include Athens, Corinth, Delos, and Oropus-Amphiareion. With one exception, every dial was constructed for use in the latitude of its provenance. The one exception is the horizontal dial (4005G) still in situ in the garden of the House of the Figured Capitals in Pompeii (40°45') which was designed for use at a much higher latitude. Most datable examples belong to the first and second centuries of our era. The earliest preserved planar dial is probably the horizontal one (4001G) discovered in a private house on Delos which derives from the second century B.C. If the vertical dials on the eight sides of Andronicus's Tower of Winds in Athens were engraved at the time the tower was built, they belong to the first century B.C. The three above-mentioned planar dials are among the few that have been found in their original position. One other, engraved in the top of a marble table, was found near the ruins of a small temple to Jupiter in Aquileia. The horizontal dial from Wiesbaden (4014G) was found in the ruins of a Roman bath. Many planar dials were found having been reused in building and paving. At least one example served as a gravestone for a Christian burial (4008G); the reverse of a second example (4010) was inscribed as a game-board.

On the whole, the planar dials from Graeco-Roman antiquity seem to be the most accurate in the corpus. Calendrical day curves and hour lines on most examples have been well placed according to the latitude of design. Only the horizontal dial from Pompeii mentioned above seems to lack internal consistency; and seven south-facing vertical dials have inaccurate semicircular networks as described in chapter 2, section 3.b. At least three day curves have been engraved on each accurate example. The equinox is always a straight line as it should be. On some examples solstice curves are approximated by broken lines (see 4012 and 4014G). On three examples, 4007, 4010, and 5021, seven day curves serve to mark the entrance of the sun into each sign of the zodiac. On 5021, all day curves except that for summer solstice are approximated by straight lines. All lines connecting hour points are straight. Comparison of the relative positions of the noon shadow on some horizontal dials shows that the gnomon was not placed vertically in the preserved gnomon hole, but was inclined toward the north point (see 4001G, 4002G, 4004G, 4005G). Gnomons on deviating dials were not fixed perpendicularly to the dial face (see 5001G and 5014G). Of course, on Graeco-Roman dials the position of the foot of the gnomon was not important because the shadow of the gnomon point marked the time.

All but five of the planar dials are of marble; the exceptions are of limestone. Traces of bronze or iron gnomons remain in lead filled gnomon holes. The solstitial hour points are marked by "dots" on the horizontal dials 4004G and 4005G from Pompeii. On many other planar dials, the broken solstice curves and straight hour lines seem to have been drawn by connecting such solstitial hour points by straight line segments. Based on this evidence and the established fact that most gnomons were not fixed perpendicular to the dial face, it is possible to devise a construction procedure for planar dials. The dialler undoubtedly marked the projection of the gnomon point on the dial's flat surface and noted the perpendicular height of the gnomon. He then would have located the noon (meridian) line, BN (see figure 40), possibly in the manner described by

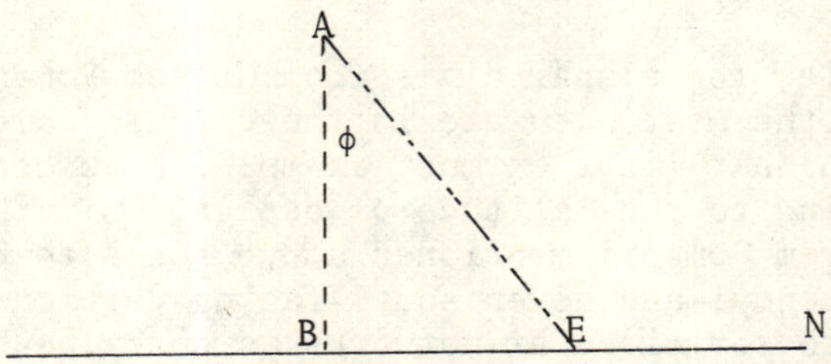

Figure 40

Vitruvius (1.6). The equinoctial line, E, could be engraved perpendicular to the noon line in its proper position, taking into consideration the perpendicular height of the gnomon and the intended latitude. The hour points at the solstices could be located using the analemmas of Vitruvius or Ptolemy (see Appendix). Connection of hour points (usually by broken lines) for a day forms the day curves; connection (with straight lines) of hour points for one seasonal hour forms the hour lines. Once the network had been engraved, the dialler could fix the position of the gnomon so that its point coincided with the gnomon point assumed in the analemma construction.

A curious triangular figure divides the dial face of dial 4001G from Delos (see figure 41). It is formed partly by two lines radiating from the point of intersection of noon line and winter solstice curve and partly by Greek inscriptions which read (right) ποῦ χρόνος πάσης ἡμέρας λοιπός and (left) ποῦ χρόνος πάσης ἡμέρας παρήκει. The lines meet the meridian line at angles $\alpha = 25°$ and $\beta = 24°$.

The inscription and the arrangement of the lines suggest that the triangle marks the increase in length of daylight from a minimum at winter solstice to a maximum at summer solstice. The lines seem to have been placed so that the part of each day curve arc outside the triangle is equal to daylight at winter solstice leaving the part of the day

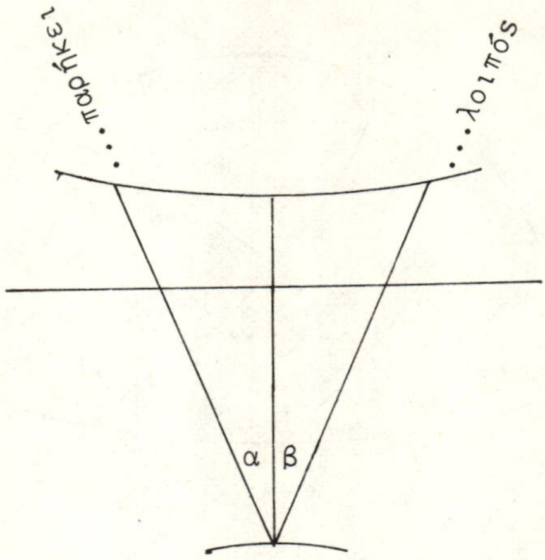

Figure 41

curve inside the triangle to mark the difference in length of daylight. This interpretation of the figure can, in fact, be justified by comparing the triangle on the Delian dial with the figure which would in theory be formed by lines marking an increase in length of daylight during the year.

Consider a horizontal dial face tangent to a celestial sphere with center at the dial's gnomon point (figure 42). The circular arcs traced on the sphere by the sun's path at solstices and equinox--SS', EE', WW'--are exactly halved by the great circle of the meridian, PAQ. The great circle of the horizon, AB, divides the circles into arcs of daylight and night, intersecting the three day circles in the figure in points D, H, and K. Let the great circle PDQ intersect the winter solstice circle at the horizon; and let it cut all the parallel circles at D, F, and G so that the arcs between PDQ and the meridian subtend angles at the earth's axis, PQ, equal to half the length of winter daylight in degrees, τ_W. Now, suppose that the relative positions of the circles on the sphere remain fixed but that the whole configuration is rotated about the earth's axis an amount equal to τ_W until PDQ lies in the meridian plane (see figure 43). Consider another great circle in the plane of the horizon A'D'B', intersecting the day circles in the figure in points D', H', and K'. Arc DD' in figure 43 is equal to arc WD in figure 42; further, HH' = EF and KK' = SG. The

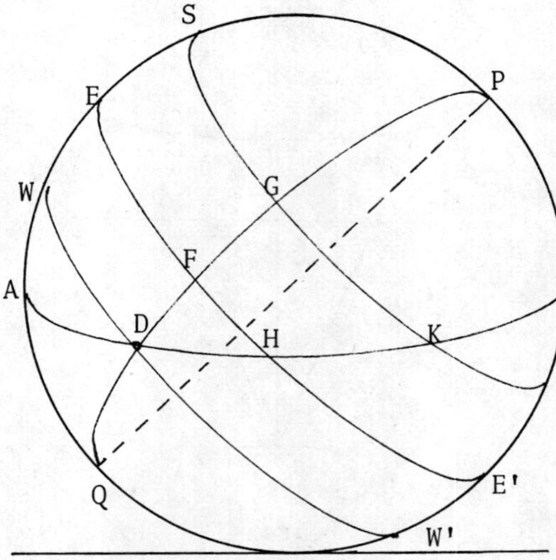

Figure 42

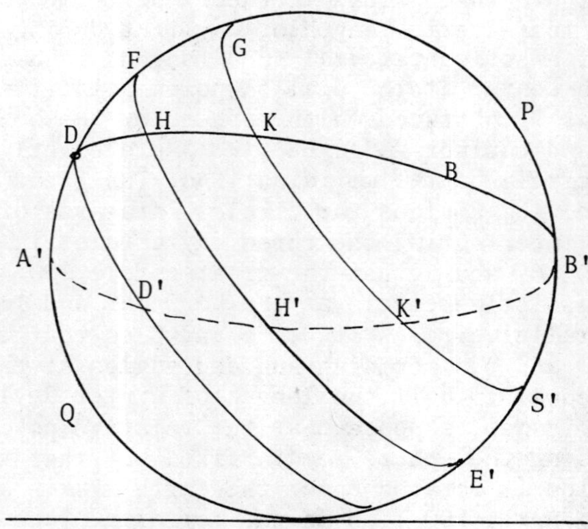

Figure 43

arc of great circle DB between the solstitial circles marks the difference in length of daylight measured from the meridian. The plane through great circle DB intersects the meridian plane in the line traced by a ray of the noon sun at winter solstice passing through the gnomon point. In figure 44, this line is OW. OWE is the meridian plane; OWP

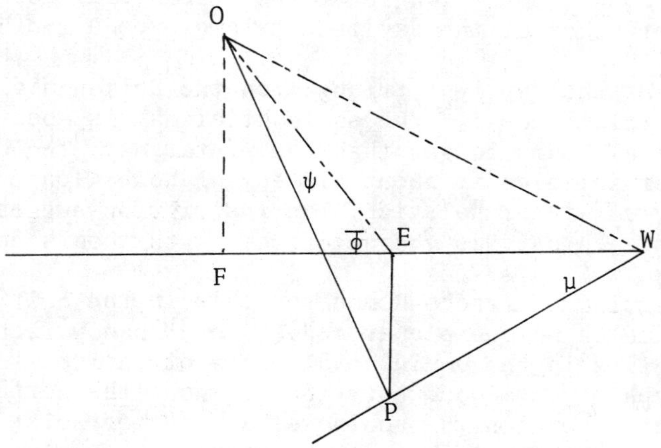

Figure 44

is the plane passing through the great circle DB. The dihedral angle formed by these planes measured in the equinoctial plane, OEP, is equal to half the difference in degrees, ψ, between winter daylight and equinoctial daylight at the latitude under consideration. The dihedral angle formed by these planes measured in the plane of the horizontal dial face is μ. The angle μ can be found by solving triangle WEP, where WE is the distance along the meridian line between equinox and winter solstice, and EP can be found from triangle OEP in the equinoctial plane:

$$\tan \mu = \frac{EP}{WE} .$$

In triangle OEP, $EP = \tan \psi \cdot OE$. In triangle OFE (in the meridian plane) $OE = \frac{OF}{\cos \phi}$. Thus,

$$\mu = \tan^{-1} \frac{OF \cdot \tan \psi}{\cos \phi \cdot WE} .$$

The value of μ computed for the intended latitude, perpendicular gnomon height, and distance WE of the Delian dial is 23°31'. (For latitudes between thirty-five and forty-five degrees, μ happens to be very close to the obliquity of the ecliptic.) The computed value of μ closely approximates the measurement of the angle β and the less accurate measurement of α, so that the triangle on the Delian dial does have the characteristics of the daylight triangle required by theory.

A glance at the daylight triangle on the Delian dial shows that at latitude 39° the equinoctial day is about 2.8 equinoctial hours longer than the winter day; the day of the summer solstice is about 4.8 summer hours longer than the day of winter solstice. The inscription suggests that perhaps each day was characterized by the hours enclosed in the triangle.

The collection of Graeco-Roman sundials in the British Museum includes a unique planar model whose shadow-receiving surface lies in the equinoctial plane of latitude 32°. The photograph of catalogue entry 5023G shows the north face, effective between the equinoxes and summer solstice. All the daylight arcs visible in the photo are greater than 180°. The south face marks the seasonal hours when the sun is south of the celestial equator in fall and winter. Two side dials with faces parallel to the meridian plane mark the equinoctial hours before and after noon. The same gnomon hole with cross section perpendicular to the equinoctial plane once held gnomons for both the north and south face. Unfortunately the museum has no record of where this dial was found. It is practically impossible to determine its date. No other equinoctial plane dials are known from antiquity.

Some geometrically inscribed flat stones have been mistakenly identified as planar sundials. In particular, these mistaken identities include stones inscribed with sectioned circles which have been found paving the streets of many Greek and Roman cities. An example from the front step of a house along the Decumanus Maximus in Ostia Antica is shown in plate 1, page 128. These engraved circles do exhibit the characteristics of the wind diagrams described by Vitruvius, or of a gameboard mentioned by Ovid, but they have nothing in common with accurately constructed planar sundials.

Cylindrical Sundials--6001-6002 and 7001-7007

The cylindrical dial with axis parallel to the axis of the earth found near Constanta, Rumania (6002G), was discussed in section 4.a of chapter 2. Two other cylindrical dials in the catalogue, from Alexandria (3086G) and Rome (3080G), have some of the properties of the Constanta dial. However, latitudes computed from measurements on the dials differ so much from the latitudes of provenance, that I suspect a dialler's error in marking either WE or ES along the meridian line.

Examples of Graeco-Roman sundials having cylindrical surfaces with axis perpendicular to the horizon have been found in Italy (7004G to 7007), in Morocco (6001), and possibly in Athens (5001G). All the preserved vertical cylinders, except the Moroccan example, are parts of dials having more than one face. Two of these cylinder combinations are remarkably similar (7004G and 7005G), and one was almost certainly modeled after the other. Three vertical cylindrical surfaces are combined in a single dial now in the Vatican Museum (7006G). One is a right circular cylinder. The other two are parts of a transposed cylindrical surface having a noncircular horizontal plane section. The curious configuration of lines on the south-facing outer surface of the reservoir chamber of the Tower of Winds may be part of a convex cylindrical dial. A test by Derek J. de Solla Price has shown that the lines would fit into a network of day curves engraved to correspond to southeast and southwest pointing gnomons. Unfortunately, not enough lines are preserved to confirm or deny the conjecture of this dial face.

Inscription

The corpus of Greek and Roman sundials contains forty-six inscribed examples. Thirty-two of these inscriptions are Greek; nine are in Latin; two are Nabatean; one is Oscan; one is Phoenician; one is neo-Punic. At least eight other inscriptions (seven Greek, one Latin) have been found on stones which were probably once parts of sundials. (These last eight inscriptions are catalogue numbers 8001-8008.) In addition to providing a means of dating the sundials, some inscriptions give names of makers and names of timekeepers. They also indicate the astronomical terminology used by diallers. A complete list of all inscribed dials is included at the end of this section.

Spaces between hour lines are marked with Greek letters on only four of the catalogue entries. These include two vertical planar dials (5003G and 5018G), one spherical dial (1055G), and one conical dial (3086G). The spherical dial was found at Ephesus and dates from the third century A.D. Very likely the other examples are also from this period or later, as the marking of hours with letters is unknown except on south-facing vertical dials from the seventh, eighth, and ninth centuries (see note 27, chapter 2, for Byzantine dials with lettered hours). The third, sixth, and ninth hours of the day are marked with special symbols on eight dials (1016G, 1022G, 1056G, 2023G, 3022, 3067G, 3076, and 3080G). The noon line of a horizontal dial found at Pompeii (4007) is inscribed μεσημβρία.

Day curves are identified in Greek or Latin on fifteen sundials--six horizontal, five spherical, four conical. The summer solstice is variously designated: τροπαὶ θεριναί (1001, 1072G, 4001G, 4007); θερινὴ τροπή (1073, 3047, 7001G); θερινή (3050G, 3058G); θερινός (1074, 3060G); ὧραι θεριναί (4009); *solstitialis* (4010, 5021). The equinox is inscribed: ἰσημερία (1001, 1072G, 1073, 4001G, 4007, 7001G); ἰσημερινὴ τροπή (3047); ἰσημερινή (3050G, 3058G, 4008G); ἰσημερινός (1074, 1075, 3060G); ὧραι ἰσημεριναί (4009); *aequinoctialis vernalis* (4010); *aequinoctialis* (5021). The winter solstice is marked: τροπαὶ χειμεριναί (4001G, 4007); τροπαί (1072G); χειμερινὴ τροπή (3047); τροπὴ χειμερινή (7001G); χειμερινή (3058G); χειμερινός (1074, 3060G); ὧραι χειμεριναί (4009); *brumalis* (4010, 5021). On two dials, a day curve near the summer solstice is inscribed κύων ἐκφανής. On a fragment of spherical dial from Delos, a day curve near the summer solstice is marked ἑῴα ἐπιτολή. A seasonal indicator, marking the morning setting of the Pleiades and the beginning of winter, between the winter solstice and the equinox on the roofed spherical dial from Tenos (7001G) reads: πλειάδων δύσις/ χειμῶνος ἀρχή. On this same dial a day curve between equinox and summer solstice is marked: πλειάς ἐκφανής/ θέρους ἀρχή, to signify the morning rising of the Pleiades and the beginning of summer.

On four dials (three Greek: 1068G, 4007, 7002G; one Latin: 4010), the day curves are marked with complete or abbreviated names of the signs of the zodiac: κρεῖος (κριός, 7002G), ταῦρος, δίδυμοι, καρκίνος, λέων, παρθένος, ζυγός, σκορπίος, τοξότης, αἰγόκερως, ὑδροχόος, ἰχθύες. *Aries*, *Taurus*, *Gemini*. Names of months are inscribed on three dials: Ιανουάριος,

Φεβράριος, Μάρτιος, Ἀπρίλιος, Μάϊος, Ἰούνιος, Ἰούλιος, Αὔγοξστος, Σεπτέμβριος, Ὀκτῶβριος, Νοέμβριος, Δεκέμβριος (1068G); Ἀπελλαῖ(ος), Δεῖος, Ὑπερβερε(ταῖος), [Γ]ορπιαῖος, Λῶος, [Π]άνημος (1074); *Januarius, Februarius, Martius, Aprilis, Maius, Junius* (5021).

Five catalogued sundials are combined with inscribed wind roses giving the following Greek and Latin wind names:

4002G	4008G	4009	4010	5001	
Eurus	εὖρος			εὖρος	SE
Auster	εὐρόνοτος	λιβόνοτος		νότος	S
Africus	[νό]τος	λίψ	*Africus*	λίψ	SW
Faonius			*Favonius*	ζέφυρος	W
Aquilo		ἀπαρκία[s]		σκίρων	NW
Septentrio		βορέας		βορέας	N
Boreas	καικίας	καικίας		καικίας	NE
Desolinus	ἀπηλιώτης	ἀφηλιώτ[ης]		ἀπηλιώτης	E

The names of dial makers are inscribed on six sundials:

Θεμισταγόρας Μενίσκου Ἀλεξανδρεύς (3049G)
Φαῖδρος Ζωίλου Παιανιεύς (5002G)
Ἀνδρόνικος Κυρρήστης (7001G)
Νεικομήδης (1004)
M Antistius Euporus (4002G)
L Sepunius and *M Herrennius* (8007).

Dedicated sundials are designated in inscriptions by the terms: ὡρολόγιον (8003, 8004, 8006); γνώμων (8005); ὡροσκόπιον (3055G). The dials are dedicated to: M Αὐρήλιος Σεουήρος Ἀντωνεΐνος and Ἰουλία, Caracalla and his mother Julia Domna (1055G); Σεβαστός, the Emperor (3004G); Βασιλεύς Πτολεμαῖος, King Ptolemy (4049G); *Sol Invictus*, Augustus[5] (1063G); the Θεοὶ Μεγάλοι (3044G and 8004); Ἀπόλλων Διδυμεύς (8002); Milk'ashtart (3094G); and the Ἐννέα Μοῦσαι (1004). Other inscriptions mention donors' names (1004, 1041, 1067G, 2017G, 3021G, 3071G, 3094G, 3095, 3096, 4001, 7002G). A curious complete, but incomprehensible, inscription appears on the base of a dial from Delos (3025G).

List of Inscriptions

Number	*Present Location*	*Inscription*
1001	Delos?	Greek: days
1004	Kavalla	Greek: hours, dedication
1041	Tripoli	Neo-Punic: dedication
1044	Bucharest	Greek: days, daylight triangle

Number	*Present Location*	*Inscription*
1055G	Selçuk	Greek: hours, dedication
1063G	Zagreb	Latin: dedication
1066G	Paris	Nabatean: dedication
1067G	Stuttgart	Latin: dedication
1068G	Vatican	Greek: months, zodiac
1072G	Delos	Greek: days
1073	Kea?	Greek: days
1074	Palmyra?	Greek: days, months
1075	Samothrace	Greek: days
2017G	Volterra	Latin: dedication
3004G	Athens	Greek: dedication
3021G	Oropus-Amphiareion	Greek: dedication
3025G	Delos	Greek: dedication?
3044G	Athens	Greek: dedication
3047	Samos	Greek: days
3049G	Paris	Greek: dedication, maker
3050G	Aphrodisias	Greek: days
3051G	Claros	Greek: dedication
3055G	Berlin	Greek: dedication
3058G	Vienna	Greek: days
3060G	Izmir	Greek: days
3071G	Naples	Oscan: dedication
3086G	London	Greek: hours
3094G	Paris	Phoenician: dedication
3095	Mada'in Salih	Nabatean: dedication
3096	Palmyra	Greek, Palmyrenian: dedication
3098	Auriol	Latin: dedication
4001G	Delos	Greek: days, daylight triangle
4002G	Aquileia	Latin: winds, maker
4007	Pompeii?	Greek: noon, days, zodiac
4008G	Vatican	Greek: winds, equinox
4009	Rome	Greek: winds, days
4010	Rome?	Latin: winds, days, zodiac
4011	Tivoli?	Latin: dedication
5001	Athens	Greek: winds
5002G	London	Greek: maker
5003G	Athens	Greek: hours
5018G	Syracuse	Greek: hours
5021	Tunis?	Latin: days, months
7001G	Tenos	Greek: dedication, days
7002G	Nafplion	Greek: zodiac, dedication
7007	Rome?	Latin: calendar

Decorative Details

Planar sundials from Graeco-Roman antiquity are undecorated except for traces of red coloring in the hour lines and day curves of two vertical examples (5019G, 5023). Similar red coloring decorates the day curves on four conical dials (3023G, 3024G, 3031G, 3092) and five spherical dials (1002G, 1023G, 1024G, 1026G, 1027G). Hour lines are emphasized with red dye on two conical dials (3024G, 3092) and on five spherical dials (1023G to 1027G). A black dye colors the hour lines on one conical dial from Delos (3023G).

Relief designs and decorative moldings were often added to nonplanar sundials. On some, a simple horizontal cylindrical molding is the only decoration (1005G, 1024G, 1032G, 3061G, 3084G). The bases of many other examples are flanked by stylized legs (1016G, 1021G, 1036G, 3001G, 3004G, 3006G to 3008G, 3010G, 3012G to 3014G, 3019G, 3021G, 3023G, 3024G, 3025G, 3028G, 3032, 3033, 3042, 3047, 3052, 3065G, 3066G, 3074G, 3095). On a few spherical dials and many more conical dials these flanking legs are lion paws (1028G, 2017G, 2020, 2021G, 2023G, 3003G, 3005G, 3011G, 3015G, 3017G, 3018G, 3020G, 3026G, 3027G, 3056G, 3057G, 3060G, 3064G, 3068, 3071, 3073, 3079, 3101, 3108G). The lion-paw motif is continued onto the side of some conical examples and embellished with honeysuckle vines, curling tendrils, and rosettes (3011G, 3016G, 3018G, 3026G, 3027G, 3064G, 3071G, 3079G). Two griffon feet decorate the base of a conical dial from Pergamon (3053G). On a conical dial in the Chios Museum (3041) and a spherical dial in the British Museum (1056G), the lion paws are surmounted by lion heads.

Small rosettes occasionally flank the dial face either in the upper or lower corners of the oblique front (1022G, 1037G, 1059, 2003G, 2007G, 2017G, 2020, 2022, 2023G, 3022). On three of these dials a horizontal crescent has been engraved between the rosettes below the noon line (1037G, 1059, 3022). On one conical dial (3076) the crescent is flanked by two geometrical figures instead of rosettes. Similar figures appear without the crescent on the oblique front of two other conical dials (3072G, 3085G). Deonna[6] interprets this figure as a symbolic carpenter's square. He does not mention that in Euclid 2, definition 2, this figure is called a γνώμων, which would make it an appropriate decoration for a sundial.

Three sundials in the corpus have bases shaped like ships (1072G, 2001G, 3107G). Two dolphins decorate the base of the roofed sphere from Tenos (7001G). The outer surfaces of two roofed spherical dials from Aquileia (2006G, 2011G) are shaped like flowers complete with calyx and stem. One of these blossoms is supported by the statue of a woman. Supporting figures are carved on the bases of many other dials including 1034, 1048, 1049, 1051G, 3045, 3054G, 3099, 3106G, and 6002G. Similar reliefs decorate two small Roman sundials with more than one dial face (7004G, 7005G). Fine profiles of Athena and Dionysos and a head of Helios were carved into the sides and base of a conical dial (3006G) now in the Pergamon Museum in Berlin. The most elaborate relief design on the base of a sundial is that on a spherical dial in Silistra, Bulgaria (1054G). It shows a youthful figure playing a harp in a garden surrounded by various animals.

Delos and Pompeii

The ancient sites of Delos and Pompeii have proven particularly fruitful sources of archeological material. For different reasons both sites were abandoned in antiquity and virtually ignored until modern times. After Mithradates VI and his pirate allies attacked for the last time in 89 B.C., the once prosperous island of Delos was never repopulated, and the ancient rubble left behind was never reused to build later cities. The ruins of Pompeii lay untouched beneath ashes and lava for seventeen centuries. Excavations at both these sites have yielded relatively large numbers of sundials.[7] The dials found at Delos date from a slightly earlier period than those found at Pompeii. Both collections provide valuable information about the skills of Hellenistic diallers in constructing scientifically designed instruments which had become such widely used timekeepers that many examples could be found in one city.

An excellent descriptive survey of the twenty-five dials which have been found by French archeologists on the island of Delos has been published in volume 18 of the *Exploration Archéologique de Délos*, pp. 187-95. The dials have typical hemispherical, conical, and planar shadow-receiving surfaces. Only one is known to have been found in a private house (4001G). Most were found in the numerous public areas, near temples and in marketplaces. The distribution is not really evidence for a lack of privately owned dials, as most of the private houses at the site have not yet been excavated. One unfinished example (1064) was found in the agora of the Italians, thus identifying at least one production center.

Analysis of the measurements of Delian sundials shows that most were made to be used on Delos or at locations with the same latitude. A single ornate conical dial seems to have been designed for use at a much lower latitude (3031G). Perhaps it was brought from Egypt as a votive offering to Isis,[8] or it may have been intended for export. Computed parameters are a witness to the general quality of construction of the preserved dials. Most examples have conical faces. Day curves have been accurately placed and occasionally identified with inscription, so that the dials function as calendars as well as clocks. The Delian finds serve as excellent evidence of the quality of workmanship and variety of design which can be attributed to the early Hellenistic dialler.

The ability of the Delian dialler's Pompeian counterpart is well illustrated by a collection of thirty-five sundials which have been recovered from the ruins of Pompeii by Italian archeologists. The earliest datable example of their work (3071G) was found near the Stabian Baths. It has a carefully divided conical face and an Oscan inscription that dates from the second century B.C. The remaining uninscribed dials have been found in later contexts and are presented together for the first time in this catalogue. Many Pompeian dials were found in private homes of the Augustan or Imperial period and remain in situ. Dials which have been found in public areas, like the Oscan cone, have since been removed to study collections or museums.

Pompeian diallers seem to have preferred a cut spherical dial face. The preserved examples were cut for latitudes between 31° and 43°45'. The intended latitudes computed for the few conical dials found at Pompeii also vary from the known latitude of that site, 40°45'. However, the intended latitude of the early Oscan cone is correct for Pompeii. Two sundials (3072G, 4005) found in the House of the Figured Capitals seem to have been made for use at a latitude near 50°. One of these dials has been engraved on the horizontal section of a column firmly planted in the garden of the house. It does not appear to have been intended for export and so can be considered the product of a careless dialler. The fragment of a roofed sphere (2019G), also made for use at a high latitude, may be another example of his work.

In general, the dials found at Pompeii are not of the

same fine quality as those from Delos, despite similarities in design. Many are made with less precision from less expensive materials (tufa covered with stucco instead of marble). A larger demand for personal timekeepers which marked the seasonal hours may have encouraged hurried, careless production. Pompeian diallers seem to have sacrificed calendrical precision in favor of simplicity of construction. They engraved three parallel day curves on their dials, but located these curves without reference to the solstices or equinox. Still, the Pompeian dials mark the seasonal hours well and give an excellent indication of the role of the sundial as timekeeper in the first century A.D.

Notes to Text

Chapter 1

1 See D. J. Price, *Gears from the Greeks, Transactions of the American Philosophical Society*, vol. 64, pt. 7 (1974); D. J. Price, "Portable Sundials in Antiquity," *Centaurus* 14 (1969): 242-66.

2 Hermann Diels, *Antike Technik*, Leipzig, 1920; Joseph Drecker, *Die Theorie der Sonnenuhren*, Berlin, 1925.

3 Ptolemy records the ancient name of these hours in *Almagest* 4.11 and 7.3: ὧραι καιρικαί.

4 Scholars presume that Vitruvius composed *De architectura* before 27 B.C. since he never refers to Octavian by the official title, Augustus, conferred on him in that year.

5 The relevant passage from Vitruvius book IX, chapter 8, is reproduced from Granger's edition (New York, 1931) at the beginning of chapter 3.

6 The analemma of Vitruvius is described in the Appendix.

7 πόλον μὲν γὰρ καὶ γνώμονα καὶ τὰ δυώδεκα μέρεα τῆς ἡμέρης παρὰ Βαβυλωνίων ἔμαθον οἱ Ἕλληνες.

8 O. Neugebauer and R. A. Parker, *Egyptian Astronomical Texts I: The Early Decans*, London, 1960, pp. 116-21. In an earlier publication ("Some Fundamental Concepts in Ancient Astronomy," in *Studies in the History of Science*, 1941, p. 28) Neugebauer refers to "gnomon tables" in the *mul APIN* as evidence that the Babylonians were familiar with primitive sundials although no examples survive. E. F. Weider discusses the text in question in "Ein Babylonisches Kompendium der Himmelskunde," *American Journal of Semitic Languages and Literature* 40(1924): 198.

9 In his article "Greek Timekeeping," *The Classical Review* 54 (1940), pp. 69 f., J. E. Powell has characterized the whole phrase in Herodotus as an Alexandrine interpolation. D. S. Robertson argues against this criticism on pages 180-82 in the same volume of *The Classical Review*.

10 In referring to the use of the word in Aristophanes, Pollux (9.46) takes πόλος to mean ὡρολόγιον.

11 In his *Natural History* (2.187), Pliny (first century A.D.) attributes the same activities to Anaximenes of Miletus, a student of Anaximander. Kirk and Raven conclude that Pliny confused the two names (*The Presocratic Philosophers*, Cambridge, 1957, p. 103). However, his account lends credence to the report that some kind of scientific instrument was erected in Sparta in the sixth century B.C. In reporting on the activities of Anaximander, Kirk and Raven (p. 102) contend that the mysterious phrase ἐπὶ τῶν σκιοθήρων refers to "a prominence in Sparta later known as 'the sundials' from the gnomon or gnomons that existed there."

12 D. W. S. Hunt, "An Archeological Survey of the Classical Antiquities of the Island of Chios Carried Out Between the Months of March and July, 1938," *The Annual of the British School at Athens*, no. 41, session 1940-45, pp. 41-42. Hunt's stone is reddish marble inscribed with a long straight meridian line divided by seven perpendicular hatch marks identified by Greek letters:

> χει(μερι)νὸς τροπικός, αἰ(γόκερως) το(ξότης),
> ὑδρ(οχόος) σκορ(πίος), ἰχ(θύες) ζυ(γός), ἰσημερινή,
> κρι(ός) παρ(θένος), ταῦ(ρος) λέ(ων), δίδυ(μοι)
> καρ(κίνος), τροπικὸς θε(ρι)νός.

Apparently no hole for the base of a shadow-caster was visible, but it would be possible to determine its position from the relative distances on the engraved line.

13 The entire relevant passage from Diogenes (2.1) reads:

> Εὗρεν [sc. Ἀναξίμανδρος] καὶ γνώμονα πρῶτος καὶ ἔστησεν ἐπὶ τῶν σκιοθήρων ἐν Λακεδαίμονι, καθά φησι Φαβωρῖνος ἐν Παντοδαπῇ ἱστορίᾳ τροπάς τε καὶ ἰσημερίας σημαίνοντα, καὶ ὡροσκοπεῖα κατεσκεύασε.

14 Plutarch (*Dion* 29.2) and a scholiast on Aristophanes (*Aves* 997) each refer to an apparently solar-related instrument, ἡλιοτροπεῖον, in use at Syracuse and Athens late in the fifth century B.C. No description of these ἡλιοτροπεῖα survives, but the name suggests that the instrument at least marked the seasons. Both objects were placed in prominent open air locations. The Athenian model was erected by the astronomer Meton.

15 In *Ecclesiazusae* 615f, two Athenian farmers, husband and wife, discuss the time of dinner in terms of a shadow length of ten feet. According to Athenaeus (*Deipnosophistae* 1.8c.c and b.243), Eubulus and Menander both told the story of a parasite invited to come to dinner when his shadow was a certain length (twenty feet--Eubulus, twelve

feet--Menander). In both versions the eager guest arrives too early by observing either a moonlight or early morning shadow of the same length as the afternoon shadow that the host had intended. Athenaeus retells these stories in the second century A.D. and we assume his readers appreciated the old method of marking time even if others were more common. One of Alciphron's *Letters* (1.3.4) indicates that parasites of the second century depended on the sundial (ὡρολόγιον) to mark mealtimes. For further discussion of the significance of the term στιοχεῖον as a measure see O. Neugebauer and H. B. van Hoesen, *Greek Horoscopes, Memoirs of the American Philosophical Society*, no. 48 (1959), p. 12.

16 Three useful works by Gustav Bilfinger consider the problem of timekeeping in ancient Greece and Rome. *Die Antiken Stundenangaben* (Stuttgart, 1888) is a philological discussion of the ancient hour. *Der Bürgerliche Tag* (Stuttgart, 1888) presents research on the beginning of the calendar day. *Die Zeitmesser der Antiken Völker* (Stuttgart, 1886) describes various methods used to indicate time of day.

17 *Epidemics* 7.25: ἕκτη [sc.ἡμέρη] πάλιν τὴν αὐτὴν ὥρην
περὶ πλήθουσαν ἀγορὴν
Epidemics 7.31: ἀπέθανε πρὶν ἀγορὴν λυθῆναι
from Littré's edition of the Corpus Hippocraticum (V396). See also K. Deichgräber, "Die Epidemien und das Corpus Hippocraticum," in *Abhandlungen Der Preussischen Akademie der Wissenshaften*, no. 3 (1933), especially pages 127-46.

18 To an observer on earth, changes in the distance of the sun are virtually imperceptible so that the sun seems to travel on the surface of a sphere which encircles the earth. Specification of a direction (important in naked-eye astronomy) is then a function of the choice of fixed reference points.

19 See Hiller's edition of Theon of Smyrna, *Expositio rerum mathematicarum*, p. 198, line 14.

20 Theon of Smyrna, p. 199, lines 6-8.

21 Strabo, *Geography* 2.1.12; 2.5.8, 2.5.41.

22 We depend on Strabo's *Geography* (written in 7 B.C.) for an approximation of the contents of the lost work of Eratosthenes. Beginning with a base parallel through Rhodes, Eratosthenes positioned other parallels according to stade differences and comparable solar phenomena. E. Honigmann (*Die Sieben Klimata und die* πόλεις

ἐπίσημοι, Heidelberg, 1929) sees the first scientific establishment of the *climates* in this division of the οἰκουμένη.

23 See Hipparchus *Commentary on Aratus* 1.3 (edition Manitius) and Strabo 2.5.34.

24 Ptolemy prefers measurements of shadows cast on a graduated miniature meridian where the height of the pole is the mean between zenith distances of the solstitial suns. In *Centaurus* 14 (1969), pp. 29-41, John Britton discusses the difficulties with Ptolemy's method as described in the *Almagest* 1.12. Whatever the problems, it seems certain that Ptolemy considered an observation of the position of the edge of a shadow to be more accurate than measurements involving the shadow of a gnomon point. He would doubtless have regarded any Graeco-Roman sundial as an unacceptable astronomical instrument. It is interesting to note that the use of gnomon-shadow ratio to specify terrestrial latitude continued in India until the nineteenth century. The preservation of this tradition is yet another indication of the influence of Greek thought on Indian astronomy.

25 Assuming the values 37°31' and 41°53' for the latitudes of Catania and Rome, respectively, and considering the obliquity of the ecliptic to have been 23°43' in 262 B.C., I have found that the longest daylight in the higher latitude would have been 226°24'; in the lower latitude, 219°26'. In Rome the shadow of the gnomon would have fallen on the engraved Catanian equinox ten days after the actual fall equinox. The greatest calendrical error would have occurred at the solstices: in Rome the gnomon's shadow would have fallen on the Catanian winter solstice *twice*--about thirty days before and after the actual winter solstice; the gnomon's shadow would never have reached the engraved Catanian summer solstice. The greatest error in time of day would have been only about .07 Roman summer hours, certainly an error small enough to go unnoticed. It seems probable that the calendrical error eventually inspired correction. If the dial was (unintentionally) tilted to compensate for the difference in latitude, both errors would naturally have gone unnoticed.

26 See Ptolemy *Almagest* 4.11 and 7.3.

27 Neugebauer, *The Exact Sciences in Antiquity*, Princeton, 1952, pp. 160-61.

28 Ibid., and pp. 218-20.

29 Ibid., pp. 226-27.
30 O. Neugebauer, "The Astronomical Origin of the Theory of Conic Sections," *Proceedings of the American Philosophical Society* 92 (1948): 136-38.

Chapter 2

1 By leveling the surface AB and hanging a plumb line which will intersect the oblique front plane, it is possible to form the angle ϕ (see figure 2). In taking measurements for the catalogue, I often employed this method.
2 T. S. Davies, "An Inquiry into the Geometrical Character of the Hour Lines upon the Antique Sundials," *Transactions of the Royal Society of Edinburgh* 8 (1818): 77-122, especially pp. 80-86.
3 In 1562 Frederico Commandino proved that corresponding seasonal hour points at the solstices and equinox lie on great circles on the sphere--see *De Horologiorum Descriptione* (Rome, 1562), pp. 54 ff. Christopher Clavius pointed out (*Gnomonices*, book 8, Rome, 1581) that seasonal hour points on day curves between solstices and equinox need *not* lie on these great circles. He proved this statement, but at the same time he remarked that the difference between great circle and seasonal hour line would not be noticed at latitudes below 45° (*Astrolabium*, 1593). Using formulas developed in the early 19th century by Settele (*Memoria sopra la forma della linee orarie*, Rome, 1816) and Delambre (*Histoire de l'Astronomie Ancienne*, Paris, 1817, pp. 476-85), Joseph Drecker computed (for various values of latitude and declination) the difference between the actual position of a seasonal hour point and its position on the great circle passing through corresponding hour points at the solstices and equinoxes. Drecker's table of differences (see *Theorie der Sonnenuhren*, pp. 14-16) confirms Clavius's observation.
4 See catalogue entries 1016G, 1017G, 1020G, 1022G, 1024G, 1025G, 1026G, 1031G, 1035G.
5 Refer to figure 4. Once $\phi = \alpha$ and r have been determined, it is possible to find the angles γ, ψ, and ζ from measurements of TW, WE, and ES. Let $\nu = \gamma + \psi + \zeta$; then the declination of the day curve at S is $\delta_S = \nu - \overline{\phi}$ and half daylight, τ_S, at that declination is $\cos^{-1}(-\tan\phi \cdot \tan\delta_S)$. The radius of the day curve, r_S,

is $r \cdot \cos \delta_S$ and one arc length between hour points on the day curve at S is theoretically equal to $(\tau_S/6 \cdot \pi \cdot r_S)/180°$. The value of this arc length computed from ϕ and r should equal the measured value $(VI$ to $VII)_S$ in the catalogue entry.

6 See catalogue entries 1013, 1048G, 1049, 1050G, 1053G, 1054G.

7 Refer to figure 5. $S'S = r_S = r \cdot \sin(\overline{\phi} + \varepsilon)$; $E'E = r_E = r \cdot \sin \overline{\phi}$; the "day curve" at S has arc length $\pi \cdot r_S$; the "day curve" at E has arc length $\pi \cdot r_E$. The arc length between hour lines on the curve at S, ℓ_S, is $(\pi \cdot r_S)/12$; the arc length between hour lines on the curve at E, ℓ_E, is $(\pi \cdot r_E)/12$. Now $12 \cdot \ell_E/12 \cdot \ell_S = (\pi \cdot r \cdot \sin \overline{\phi})/[\pi \cdot r \cdot \sin(\overline{\phi} + \varepsilon)]$ $= \sin \overline{\phi}/(\sin \phi \cos \varepsilon + \cos \phi \sin \varepsilon) =$ $1/(\cos \varepsilon + \cot \overline{\phi} \sin \varepsilon)$; $\tan \phi = [(\ell_E/\ell_S) - \cot \varepsilon]/\sin \varepsilon$; $\overline{\phi}$ = intended latitude. Finally $r = (12 \cdot \ell_S)/[\pi \cdot \sin(\phi + \varepsilon)]$ or $r = (12 \cdot \ell_E)/(\pi \cdot \sin \overline{\phi})$; r = gnomon length.

8 The dial is now preserved in the Archeological Museum in Bergama, Turkey. It is number 1069G in this catalogue. See Albert Rehm, "Eine Zwillingssonnenuhr aus Pergamon," *Mitteilungen des deutschen archäologischen Instituts, Athenische Abteilung* 36(1911): 251-68.

9 For a discussion of the mathematical character of the day curves when projected onto the meridian and equator planes see Drecker, *Theorie der Sonnenuhren*, pp. 34-36.

10 Given a circle of radius r and a secant which rotates about a point O on the circle and intersects the circle again in point A. The limaçon is the locus of points on OA (inside and outside the circle) always a given distance d from A. Jordanus Nemorarius called the curve the "conchoid of the circle." Etienne Pascal was the first to carefully study the curve. The equation of the curve in perpendicular coordinates originating at O is in general: $(x^2 + y^2 - 2rx)^2 = d^2 \cdot (x^2 + y^2)$.

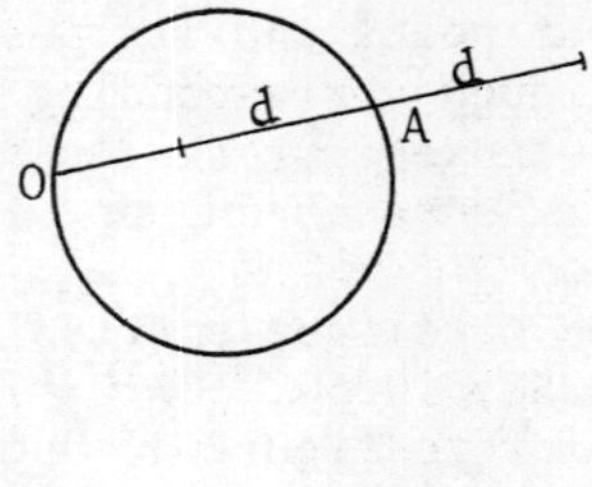

Drecker (p. 27) gives the specific equation for the curve in the equator plane of a pierced spherical dial: $(x^2 + y^2 - 2r \cdot \cos\phi \cdot \cos^2\delta \cdot x)^2 = r^2 \sin^2\phi \cdot \sin^2 2\delta(x^2 + y^2)$ where r is the radius of the spherical shadow-receiving surface, ϕ is the latitude, and δ is the solar declination. The three dimensional limaçon engraved on the interior of the pierced dial is essentially the locus of points on great circles passing through the gnomon point, at a given distance from the equator circle.

11 Rehm, *Ath. Mitt.* 36(1911): 256.

12 The digamma symbol represents the number 6, here presumably associated with the sixth hour of the day or noon.

13 Cicero *De re publica* 1.14 (21-27), Keyes's translation.

14 T. S. Davies, pp. 97-104.

15 Constantine Ionescu - Carligel presents this latter method for finding ϕ in his article, "Contributions à l'Étude des Cadrans Solaires Antiques," *Dacia* 14 (1970): 130-31.

16 As with the roofed spherical dials, it is also quite possible that the part of the conical surface between O and A was not complete but trimmed and notched in a manner similar to that pictured in figure 10. The point O may lie anywhere along VA so that S falls on VB.

17 Refer to figure 21. The nature of these curves can be determined from a consideration of the intersection of the equations of these two cones related to a coordinate system where the axis of the earth PO is the Z-axis, the intersection of the meridian plane and the plane of the equator is the X-axis, and the intersection of the prime vertical plane and the plane of the equator is the Y-axis. The equation of the sunlight cone is then $z^2 = \tan^2\delta(x^2 + y^2)$ and the equation of the conical receiving surface is $z^2 = 2r \cdot \tan\overline{\phi} \cdot z = \tan^2\overline{\phi}(x^2 = 2rx + y^2)$. The second equation is derived by a translation of axes such that $x' = x - r$ and $z' = z - r \cdot \tan\overline{\phi}$ and the cone has translated equation $z'^2 = \tan^2\overline{\phi}(x'^2 + y^2)$. Combinations of the two equations show that a nonequinoctial curve projected onto the meridian plane is a parabola. The same nonequinoctial curve projected onto the plane of the equator has the equation of Pascal's limaçon.

18 At the equinox $\delta = 0$, and the equation in the plane of the equator is $x^2 - 2rx + y^2 = 0$. In the meridian plane, the equation is $z = 0$, a straight line coinciding with the X-axis as we have defined it.

19 The attached figure shows the meridian section of a horizontal dial at latitude 66°N. Since SF is parallel

to OW', the summer solstice day curve will be a parabola. Above latitude 66°N, the summer solstice curve is elliptical until it becomes a circle at the pole. The winter solstice sun does not cast a shadow above the arctic circle. At the pole the equinoctial shadow of the gnomon *point* does not fall on the dial face. Drecker (*Theorie der Sonnenuhren*, p. 43) derives the equations of these day curves with reference to a Cartesian coordinate system with center at P.

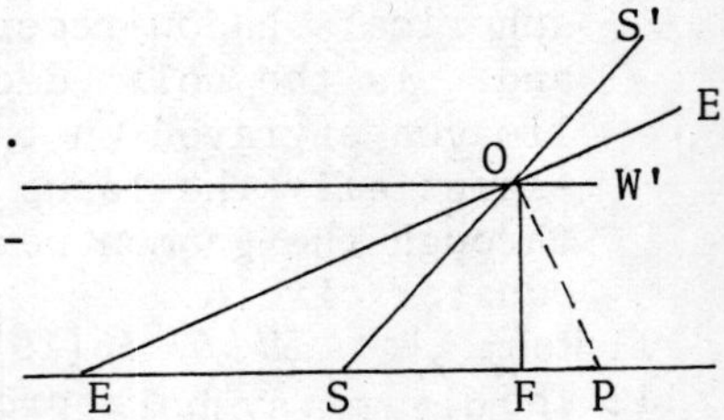

20 At sunrise and sunset, the shadow of the gnomon point does not fall on the horizontal plane. Hence the beginning of the first hour of the day and the end of the last hour of the day cannot be marked on a horizontal dial face.

21 For the equations of the nonlinear hour lines on a horizontal dial, see Drecker, *Theorie der Sonnenuhren*, pp. 43-44.

22 As was explained in note 3, Clavius was the first to note the adequacy of this approximation for latitudes below 45°.

23 On all the dials included in the catalogue, measurements of shadow length and direction were made from the center of the gnomon hole, G. If comparison of shortest noon shadow and measured distance GS shows that the center of the gnomon hole does not coincide with the projection of the gnomon point F, then the values found from formulas (3.1), (3.2), and (3.3)--i.e., FQ and angle SFQ--should be converted to the measured shadow length and direction, GQ and angle HGQ, by solving triangle GFQ.

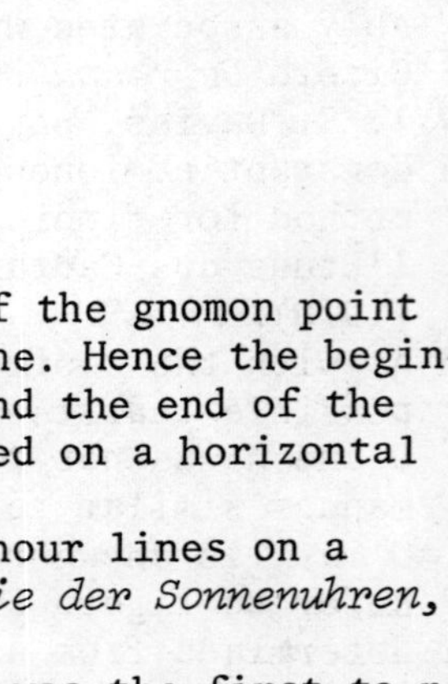

24 As the figure illustrates, at the Tropic of Cancer the summer sun casts an infinitely long shadow on a vertical dial. The trace of the shadow of the gnomon point at the winter solstice is a parabola. This winter solstice day curve becomes an ellipse and then a circle for latitudes below the torrid zone up to the equator.

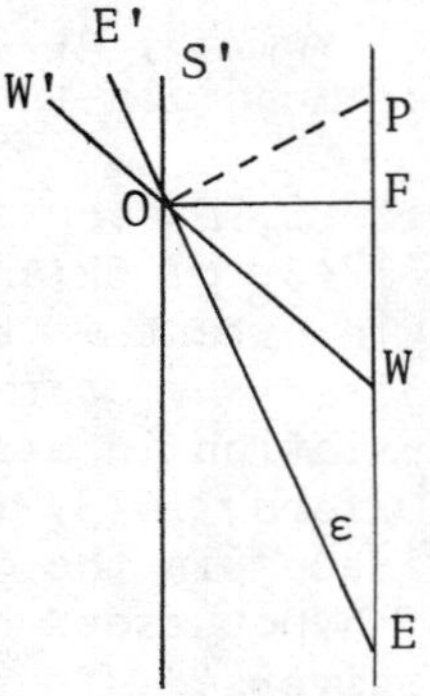

25 For azimuth greater than 90°, the sun's rays will fall on the south face of a prime vertical dial. For azimuth less than 90° the sun's rays will fall on the north face of a prime vertical dial.

26 For a north-facing dial in the plane of the prime vertical, azimuth greater than 90°, $q_V = - h \cdot \tan (z_V)$.

27 A fifth century A.D. dial from Tell Bisé (*Syria* 9 [1928], p. 80) and three Byzantine dials from Thebes, Skripù (*Byzantinische Zeitschrift* 3 [1894], p. 16) and Istanbul (G. Mendel, *Catalogue des Sculptures Grecques, Romaines et Byzantines*, vol. 2, p. 567) are all of the protractor type. The Tell Bisé example is particularly noteworthy because it is graduated within the usual twelve divisions, making possible more accurate estimations of the direction angle. For an informative discussion of the Saxon dials, see T. W. Cole, *Origin and Use of Church Scratch Dials*, London, 1935.

28 A plane parallel to the meridian always intersects the sunlight cone parallel to its axis.

29 Drecker gives the theoretical equations of the nonlinear hour lines on pp. 49-54 of *Theorie der Sonnenuhren*. Again the approximation is negligible in temperate latitudes.

Chapter 3

1 *Marcus Vitruvius Pollio, Baukunst*, translated by August Rode, Leipzig, 1796; Albert Rehm, "Horologium," *Paulys Real-Encyclopädie*, pp. 2416-28; Hermann Diels, *Antike Technik;* Joseph Drecker, *Theorie der Sonnenuhren; Vitruve, De l'Architecture*, translation and commentary by Jean Soubiran, Paris, 1969; E. Ardaillon, "Horolo-

gium," *Dictionnaire des Antiquities Greques et Romaines*, vol. 3, pt. 1, pp. 256-64; D. J. Price, "Portable Sundials in Antiquity," *Centaurus* 14 (1969): 242-66.

2 Luca Zuzzeri, *Dissertazione d'un Antico Orologio a Sole*, Venice, 1746, p. 62 and plate.

3 The single extant work by Aristarchus is his treatise, *On the Sizes and Distances of the Sun and Moon* (see text and translation in Sir Thomas Heath, *Aristarchus of Samos*, Oxford, 1913, reprinted in 1966). He is mentioned by Archimedes in the *Sandreckoner* as the author of a "book of hypotheses" containing a heliocentric theory of the universe.

4 Cleomedes *De Motu Circulari* 1.10, ed. Ziegler, p. 100. The *Suda Lexicon* places the birth of Eratosthenes in the 126th Olympiad (276-273 B.C.). He was invited to Alexandria by Ptolemy Euergetes I and he probably made his observations with the scaphe about 240 B.C.

5 See Soubiran, p. 247.

6 Photographs of an apparently independently functioning, disk-shaped meridian dial presented in catalogue entry 5024 were taken from originals preserved in the files of the Museo Nationale Romano (neg. 54820). Unfortunately the object itself was lost in the storage rooms of the museum in the summer of 1971, and I was unable to examine it. A sketch appended to Francesco Peter's *Di un Antico Orologio Solare*, Rome, 1815, p. 36, suggests that this type of dial was not unique.

7 See Ardaillon and Rehm in the articles listed in note 1.

8 Paul Tannery, *Histoire de l'Astronomie*, Paris, 1893, pp. 49, 51 ff.

9 According to the chronicle of Apollodorus given by Diogenes Laertius (i.90), Eudoxus flourished in the 103d Olympiad (368-365 B.C.). Hultsch's article on "Eudoxus," in *Paulys Real-Encyclopädie* is still the best account of what is known about Eudoxus. For additional material, see F. Lasserre, *Die Fragmente des Eudoxos*, Berlin, 1966, especially pp. 137 ff.

10 For a very useful summary of what little is known about Apollonius of Perga, see the relevant article by G. J. Toomer in vol. 1 of the *Dictionary of Scientific Biography*, pp. 179-93.

11 Diels, *Antike Technik*, p. 174.

12 "Les Cadrans Solaires Coniques," *Annales de Chimie et de Physique*, 5th ser. 6 (September, 1875), p. 60.

13 The Latin term *plinthium* derives from the Greek word πλίνθος which generally refers to a squared block of stone. However, in two inscriptions (SIG^{3}247I22 and IG IV^2 102.73 both dating from the fourth century B.C.) the term may possibly refer to a metal plate set into a threshold. Perhaps the roofed spherical dial was designated by a technical term for the metal plate set into its top surface.

14 Diels, *Antike Technik*, p. 179, note 1; W. Kubitschek, *Grundriss der Antiken Zeitrechnung*, Munich, 1928, p. 197 f; Delambre, *Histoire de l'Astronomie Ancienne*, p. 517 f.

15 See Soubiran, pp. 257-58. Soubiran's source is F. Gisinger's article on Patrocles in *Paulys Real-Encyclopädie*, vol. 18, pt. 2, pp. 2263-73.

16 See Soubiran's commentary on book 9 of *De architectura*, p. 261; Ivor Bulmer Thomas, "Dionysodorus," *Dictionary of Scientific Biography*, vol. 4, pp. 108-10.

17 See Rayet, p. 58; Drecker, p. 37; and the Loeb edition of *De architectura*, London, 1956, p. 255, n. 7. Pliny mentions the contribution of Dionysodorus of Melos in book 2, line 248 of his *Natural History*.

18 Eutocius, Commentary on book 2 of the *Sphere and the Cylinder* in *Archimedes Opera Omnia*, vol. 3, ed. Heiberg, Leipzig, 1914, pp. 152.27-160.2.

19 Eutocius, Commentary on book 1 of the *Sphere and the Cylinder*, ibid., pp. 78, 96-97.

20 Apollonius shows that even with this relaxed definition one still gets precisely the same family of curves. Apollonius, *Conics*, book 1, propositions 11, 12, and 13, in *Apollonii Pergaei quae graece exstant cum commentariis antiquis*, vol. 1, ed. Heiberg, Leipzig, 1891.

21 E. J. Dijksterhuis, *Archimedes*, Copenhagen, 1956, pp. 63-66.

22 See "Les Cadrans Solaires Coniques," pp. 61-63, for a discussion of the inscription which reads

Βασιλεῖ Πτολεμαίῳ Ἀπολλ[ώ]νιος Ἀπολλοδότου
Θεμισταγόρας Μενίσκου Ἀλεξανδρεὺς ἐποίει

23 See Soubiran, p. 266.

24 Ionescu-Carligel, "Contributions a l'Étude des Cadrans Solaires Antiques," p. 10.

25 Price, "Portable Sundials in Antiquity," pp. 242-66.

26 Derek J. de Solla Price summarizes and clarifies Vitruvius's description of these clocks in his article "Precision Instruments: To 1500," in *A History of*

Technology, edited by Charles Singer, London, 1957, pp. 601-05.

27 O. Neugebauer, "On Some Aspects of Early Greek Astronomy," *Proceedings of the American Philosophical Society* 116(1972): 243-51.

Chapter 4

1 For a detailed discussion of this "daylight triangle" refer to the section on planar sundials in this chapter, pages 80 to 84.

2 Apparently a Roman dial in the collection of Fr. J. Giry, curator of the National Museum of Euserune, Nissan-les-Euserune, France, is preserved with gnomon intact (see catalogue 1046). I have not personally examined this example but depend for my information on the notes and sketch of L. Janin. The straight gnomon indeed extends horizontally from the top back of the spherical surface.

3 Rayet, "Les Cadrans Solaires Coniques," pp. 83-85.

4 Renate Tölle, "Uhren auf Samos," in *Opus Nobile*, Wiesbaden, 1969, pp. 164-71.

5 Probably Heliogabalus. In any case, an emperor of the third century or later following the development of the cult of Sol Invictus in the Roman world.

6 W. Deonna, *Bulletin de Correspondance Hellénique* 56(1932): 470.

7 Dials from Delos:
horizontal - 4001G
vertical - 5008G to 5010
conical - 3024G to 3033
spherical - 1002 and 1070G to 1072G
unfinished - 1064

Dials from Pompeii:
horizontal - 4004G to 4007
vertical - 5015
conical - 3064G to 3073G
spherical - 1019G to 1031G and 2019G

8 At least two inscriptions concerning dedicated ὡρολογία have been found at temples of Isis on Delos. See P. Roussel, *Les cultes égyptiens à Délos*, 1915-1916, p. 177, no. 173 and p. 235, line 74.

Appendix The Analemmas of Vitruvius and Ptolemy

Vitruvius

In book 9, chapter 7 of *De architectura,* Vitruvius describes a figure, *analemmatos*, from which "many varieties and kinds of sundials" may be deduced and drawn. The figure represents an effort to avoid a three-dimensional representation of the celestial sphere by considering various circles on the sphere projected or rotated onto the meridian plane.

The analemma's relation to dialling is evident in the first step of its construction. Two perpendicular lines are drawn to represent a gnomon, AB, and its equinoctial noon shadow, BC (see figure 45). The circle with center A and radius AB represents the meridian with AC a ray of sunlight at the equinox. The line EAI parallel to BC is the projection of the horizon passing through A onto the meridian plane.

Vitruvius continues the construction of the analemma by marking off 1/15 of the meridian circle (24°) on either side of F, the point of intersection of meridian and equinoctial ray. The division points are marked H and G, with AH (the ray of the sun in summer) intersecting BC in T and the upper meridian circle in L, and AG (the ray of the sun in winter) intersecting BC extended in R and the upper meridian circle in K. LG is the projection of the parallel circle of the summer sun onto the meridian plane; KH is the projection of the parallel circle of the winter sun onto the meridian plane. LYG and KXH are the eastern halves of these parallel circles rotated into the meridian plane. PQ, passing through A and the midpoints M and O of LG and KH represents the axis of the earth. Lines VX and SY, parallel to PQ, are the intersections of the horizon and the solstitial circles rotated into the meridian plane. The horizon plane intersects the diameters LG, KH, and NF at right angles. Arc LY represents the half-daylight in summer; arc KX represents the half-daylight in winter.

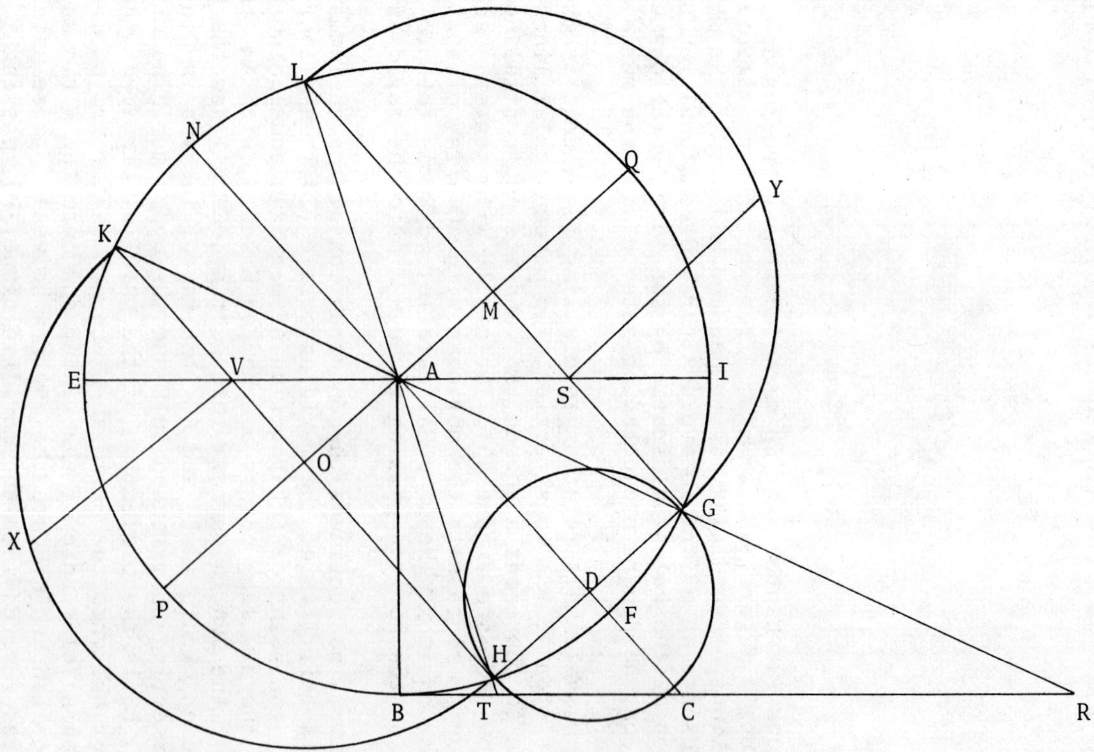

Figure 45

Instructions regarding the completion of the analemma, somewhat confusing in the original, have greatly benefited from the comments of later authors. Vitruvius directs the reader to place the center of the compasses at the point D "quo secat circinationem aequinoctialis radius," and to extend the compasses to H "quo secat circinationem aestivus radius," drawing the circle of the months with radius DH. This final addition to the analemma is called the *menaeus*. In 1581, Christopher Clavius (*Gnomonices Libri Octo*) both clarified and justified these last instructions, emphasizing that the point D is the point of intersection of equinox and chord GH. (See figure 45 for a diagram satisfying Vitruvius's instructions with the clarification of Clavius.)

Starting from Clavius's clarified definition of the *menaeus*, Joseph Drecker (*Theorie der Sonnenuhren*) gives a simple proof of the use of the circle in marking the entrance of the sun into each of the signs of the zodiac. In figure 46, LH is the projection of the circle of the ecliptic onto the meridian plane; the semicircle on LH is the ecliptic rotated into the meridian plane. The point λ is one of six points which equally divide the arc LH; α' is the projection of λ onto the diameter LH. The line through α' parallel to the equinox is the projection of the path of the sun (as it enters a sign of the zodiac) onto the meridian plane. Arc LL' equals 90°.

As Drecker shows, the same line results from proper division of the *menaeus*. Let the point α divide the *menaeus* at the point where it is intersected by a line through α' parallel to the equator. It can be shown that $LA\lambda$ is equal to $GD\alpha$. Now the $\sin G'D\alpha = D\beta/D\alpha$; $D\alpha = DH = \sin \varepsilon$, where ε is the obliquity of the ecliptic, and the radius of the meridian circle is equal to one; so $\sin G'D\alpha = D\beta/\sin \varepsilon$. But $\sin L'A\lambda = \alpha'A = \alpha'\beta'/\sin \varepsilon = D\beta/\sin \varepsilon$. Hence, $\sin L'A\lambda = \sin G'D\alpha$, $L'A\lambda = G'D\alpha$, and $LA\lambda$ is equal to $GD\alpha$, with $\alpha\alpha'$ the projection of the parallel circle of the sun at longitude $G'D\alpha$; $A\gamma$ represents a ray of sunlight at noon for solar longitude $G'D\alpha$. Thus, division of the *menaeus* according to solar longitude can lead to proper placement of the corresponding day curve on the sundial.

Vitruvius hints at the application of his analemma to the construction of sundials but it was left to later commentators to provide detailed instructions. Gustav Bilfinger (*Die Zeitmesser der Antiken Völker*, pp. 28-37) shows how the analemma may be used to construct a spherical and horizontal sundial. Drecker (*Theorie*, pp. 1-4) applied the ana-

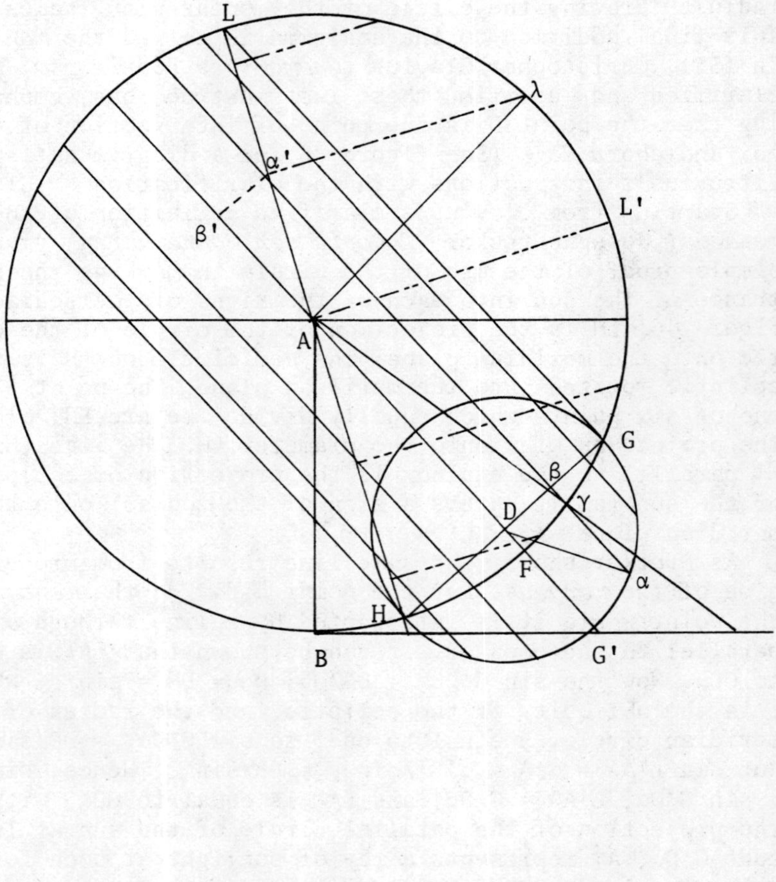

Figure 46

lemma to the construction of various plane surface and cylindrical surface sundials.

Ptolemy

Some time in the first half of the second century A.D., Ptolemy considered the angles treated "in locum gnomonicum." He presented his considerations in a brief work, Περὶ ἀναλήμματος. Parts of the Greek text of this work are preserved in *Ambrosianus L99*, a palimpsest (examined by Heiberg in Milan in the nineteenth century) dating from the seventh century. A Latin translation survives in the codex *Vaticanus Ottobonianus lat. 1850*, written in the thirteenth century. The most reliable edition of Περὶ ἀναλήμματος appeared in 1905, when Heiberg published it among Ptolemy's *Opera Astronomica Minora*, presenting both the partial Greek text from *Ambros. L99* and a complete Latin translation prepared by William of Moerbeke according to *Ottobon. lat. 1850*.

The dialler's angles with which Ptolemy is concerned in Περὶ ἀναλήμματος are those which determine solar altitude and direction. He includes them in a system of spherical coordinates which he considers an improvement on earlier attempts. Ptolemy's predecessors, "the ancients," referred solar position to four fixed great circles on the celestial sphere: the meridian, the horizon, the prime vertical, and the equator; and two movable great circles: the descensivus circle, passing through zenith, nadir, and the sun; and the horarius circle, passing through north and south points and the sun. The six angle measurements (*hectemoros*, *horarius*, *antiskios*, *descensivus*, *in plano equinoctialis*, *in plano ejus qui secundum verticem*) taken along these circles are shown in figure 47. Any two determine solar position. Ptolemy revises this ancient system by replacing the fixed equator circle with a movable hectemoros circle which passes through east and west points and the sun. The axes of the fixed circles (gnomon, meridian line, equinoctial line) are mutually orthogonal. The movable circles rotate about these axes. (See figure 48.)

Ptolemy takes acute angular measurements along these circles between the axes and a radius of the celestial sphere passing through the sun. He preserves the ancient practice of measuring the horarius angle along the horarius circle from south (or north) point to the sun and of measuring the vertical angle along the prime vertical circle from zenith (or nadir) to the point of intersection

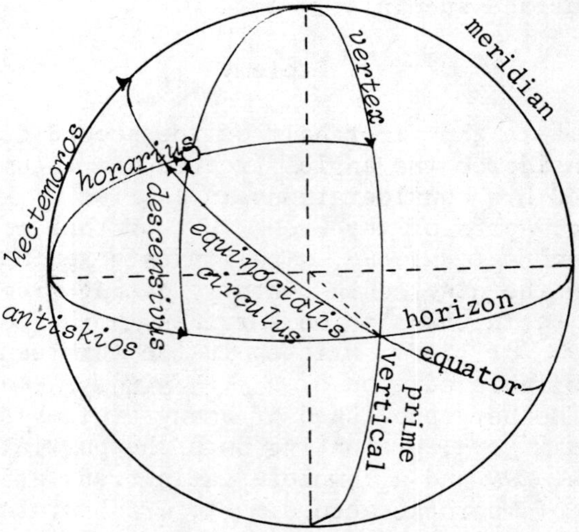

Figure 47

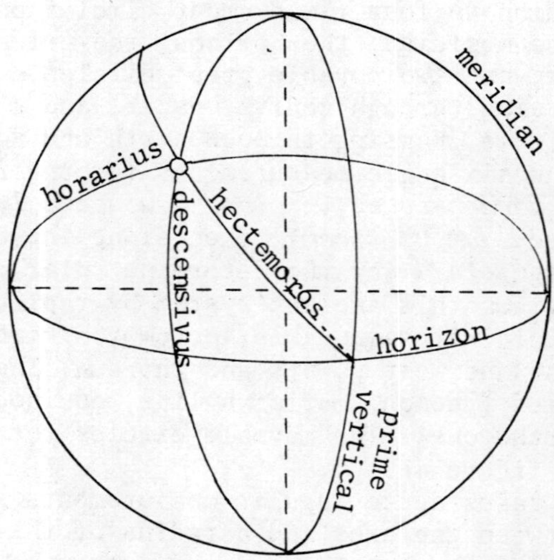

Figure 48

of horarius circle and vertical circle. He measures, as the ancients did, the angle along the meridian circle from the south (or north) point to the point of intersection of meridian and movable hectemoros circle; but he calls this the meridian angle. He reserves the name hectemoros angle for the measurement taken along the hectemoros circle from east (or west) point to the sun. He prefers to measure the descensivus angle along the descensivus circle from the zenith (or nadir) to the sun, and he measures the horizon angle along the horizon circle from the east (or west) point to the point of intersection of descensivus circle and horizon circle. Figure 49 illustrates Ptolemy's angular

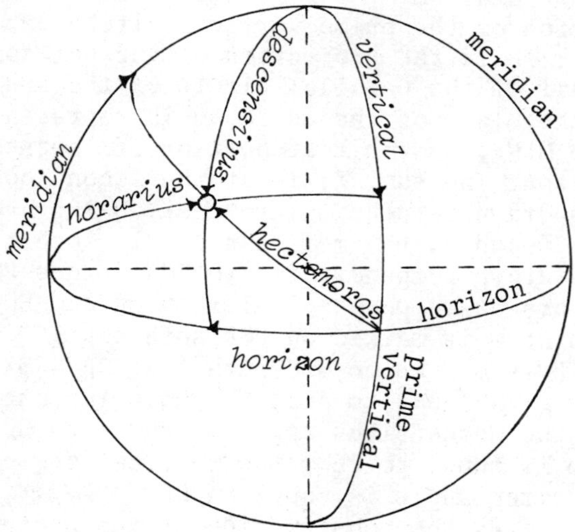

Figure 49

measurements. Again, any two determine solar position. (Ptolemy's revision of the ancient system is emphasized by a comparison of figures 47 and 49.)

Having thus described the components of his system of celestial coordinates, Ptolemy devotes the rest of the ἀνάλημμα to a discussion of the methods for determining the values of these coordinates, given the latitude of the observer, ϕ, the longitude of the sun, λ (hence, its declination, δ), and the seasonal hour, τ--all in degrees. One suggested method is trigonometric; the other, nomographic. Ptolemy could have used either method to produce the table which concludes the preserved part of his discussion. He computes hectemoros, horarius, descensivus, meridian, vertical, and horizon angles for every seasonal hour of a longest day equal to thirteen equinoctial hours; i.e., δ = 23°50', ϕ = 16°27'.

The determination of the values of Ptolemy's six coordinates is essentially a problem in spherical trigonometry and yet the whole point of his ἀνάλημμα seems to be to avoid this method of solution. By rotating and projecting parts of celestial circles onto the meridian plane, he constructs a diagram which yields the value of his angles upon application of plane trigonometry or nomography. Figure 50 illustrates the basic diagram: AGBD is the meridian circle; GD is the projection of the prime vertical circle onto the meridian plane; AB is the projection of the horizon circle. The eastern half of the parallel circle of the sun's path, rotated into the plane of the meridian is represented by the semicircle HXNK; HXN is that portion of the semicircle above the horizon. The sun, X, is at the second hour point. The line XO is drawn perpendicular to HK. Using a compass, the point Y is found on the meridian circle such that OY = OX. EΛ is drawn perpendicular to EOW. Lines POR and SOC through O are drawn perpendicular to EB and EG respectively. The point Q is marked on POR such that PQ = OX, and the point F is marked on SOC such that SF = OX. The hectemoros angle is equal to ΛEY; the horarius angle is equal to BER; the descensivus angle is equal to GEC; the meridian angle is equal to BEW; the vertical angle is equal to GEΨ; the horizon angle is equal to GEΩ.

Ptolemy justifies the construction of the hectemoros angle by considering the intersection between prime vertical and horizon (the equinoctial line), EV, and the parallel circle HXNK in their original positions perpendicular to the meridian plane. Since EX is the radius of the celestial sphere passing through the sun, and EV is on the equinoctial line, then the angle XEV satisfies Ptolemy's definition of the hectemoros angle. In triangles OEY and OEX, EX and EY are equal radii of the celestial sphere; side EO is common to both triangles; YO has been constructed equal to XO. Hence, triangle YEO is congruent to triangle XEO and angle OEX is congruent to angle OEY. Angles OEΛ and OEV are right angles, where EV is on the equinoctial line. Therefore, the complement of OEX is equal to the complement of OEY, and YEΛ equals XEV. (Since EOX is a right angle, EOY must also be a right angle; YO is parallel to EΛ and

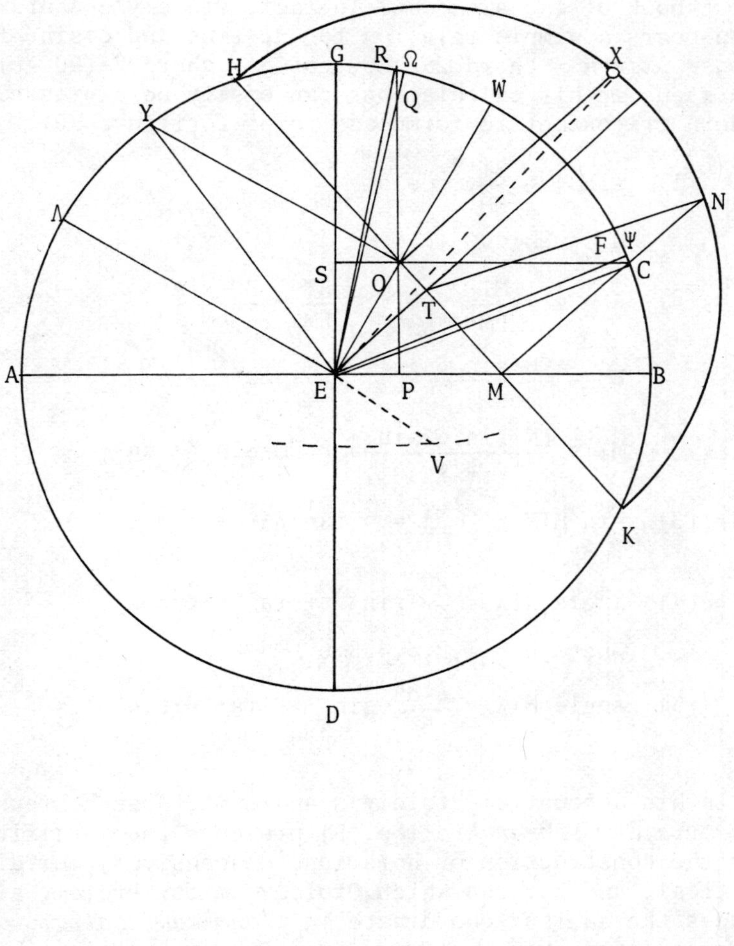

Figure 50

YEΛ = EYO.) Ptolemy apparently felt that the plane construction of the remaining coordinates had been adequately justified by earlier authors. He does not repeat their arguments,* but begins to explain how the diagram may be used to compute the angles it represents.

Ptolemy's plane trigonometric solution requires the measurement of lines in his diagram which can be related to a table of chords like that presented in chapter 11 of the first book of the *Almagest*. In fact, Ptolemy's chord function bears a simple relation to our sine and cosine functions: $\sin\alpha = \text{chord } 2\alpha/120$, $\cos\alpha = \text{chord } 2\cdot(90-\alpha)/120$. The steps in his calculations can easily be expressed in modern trigonometric formulas (refer to figure 50):

(1) $\sin\delta = \dfrac{ET}{EH}, \quad \cos\delta = \dfrac{HT}{EH}$

$HT = EH\cdot\cos\delta$

(2) $\sin\phi = \dfrac{MT}{EM}, \quad \cos\phi = \dfrac{ET}{EM} = \dfrac{EH\cdot\sin\delta}{EM}$

$EM = \dfrac{EH\cdot\sin\delta}{\cos\phi}$

$MT = \dfrac{EH\cdot\sin\delta\cdot\sin\phi}{\cos\phi} = EH\cdot\sin\delta\cdot\tan\phi$

(3) $\sin HTN = \dfrac{MT}{HT} = \tan\delta\cdot\tan\phi$

(4) $\text{angle } NTX = \dfrac{n}{6}\cdot\sin^{-1}(\tan\delta\cdot\tan\phi)$,

where $n = 1,2,3,4,5,6$

(5) $\text{angle } HTX = \dfrac{6-n}{6}\cdot\sin^{-1}(\tan\delta\cdot\tan\phi) = \tau$

*In his account of Ptolemy's analemma, Joseph Drecker (*Theorie der Sonnenuhren*, p. 5) presents the justification for the construction of horarius, descensivus, meridianus, vertical, and horizon which Ptolemy omits. Ptolemy also includes the ancient coordinate *in plano equinoctialis* in his diagram (angle MXO in figure 50) without justifying its construction. A suitable justification is provided by P. Luckey in his article, "Das Analemma von Ptolemaeus."

(6) $\sin HTX = \frac{XO}{TX} = \frac{XO}{HT} = \frac{XO}{\cos \delta \cdot EH}$

(7) $\cos HTX = \frac{OT}{TX} = \frac{OT}{HT} = \frac{OT}{\cos \delta \cdot EH}$

(8) triangle EMT is similar to triangle OPM and $\frac{EM}{MO} = \frac{MT}{PM} = \frac{ET}{OP}$

(9) $PO = \frac{MO \cdot ET}{EM} = \frac{ET(MT + TO)}{EM}$

$= \frac{EM \cdot \cos \phi\,(EH \cdot \tan \phi \cdot \sin \delta + EH \cdot \cos \delta \cdot \cos \tau)}{EM}$

$= EH \cdot \cos \phi \cdot (\tan \phi \cdot \sin \delta + \cos \delta \cdot \cos \tau)$

(10) $PM = \frac{MO \cdot MT}{EM}$

$= \frac{EM \cdot \sin \phi(EH \cdot \tan \phi \cdot \sin \delta + EH \cdot \cos \delta \cdot \cos \tau)}{EM}$

$= EH \cdot \sin \phi(\tan \phi \cdot \sin \delta + \cos \delta \cdot \cos \tau)$

(11) $\cos YE\Lambda = \cos EYO =$ cos hectemoros

$= \frac{YO}{EY} = \frac{OX}{EH} = \frac{EH \cdot \cos \delta \cdot \sin \tau}{EH} = \cos \delta \cdot \sin \tau$

(12) $\cos BER =$ cos horarius

$= \frac{EP}{ER} = \frac{EP}{EH} = \frac{EM - PM}{EH}$

$= \frac{1}{EH} [EH(\frac{\sin \delta}{\cos \phi} - \sin \phi \cdot (\tan \phi \cdot \sin \delta + \cos \delta \cdot \cos \tau))]$

$= \cos \phi \cdot \sin \delta - \cos \delta \cdot \sin \phi \cdot \cos \tau$

(13) $\cos GEC =$ cos descensivus

$= \frac{SE}{EC} = \frac{OP}{EH} = \sin \phi \cdot \sin \delta + \cos \phi \cdot \cos \delta \cdot \cos \tau$

(14) $\tan BEW =$ tan meridian

$= \frac{\sin PEO}{\cos PEO} = \frac{OP}{EP}$

$$= \frac{EH \cdot \cos\phi\,(\tan\phi \cdot \sin\delta + \cos\delta \cdot \cos\tau)}{EH \cdot \cos BER}$$

$$= \frac{\sin\phi \cdot \sin\delta + \cos\phi \cdot \cos\delta \cdot \cos\tau}{\cos\phi \cdot \sin\delta - \sin\phi \cdot \cos\delta \cdot \cos\tau}$$

(15) $\tan GE\Psi = \tan$ vertical

$$= \frac{\sin SEF}{\cos SEF} = \frac{XO}{OP} = \frac{EH \cdot \cos\delta \cdot \sin\tau}{EH \cdot \cos\phi \cdot (\tan\phi \cdot \sin\delta + \cos\delta \cdot \cos\tau)}$$

$$= \frac{\cos\delta \cdot \sin\tau}{\sin\phi \cdot \sin\delta + \cos\phi \cdot \cos\delta \cdot \cos\tau}$$

(16) $\tan GE\Omega = \tan$ horizon

$$= \frac{\sin EQP}{\cos EQP} = \frac{EP}{OX}$$

$$= \frac{EH \cdot (\cos\phi \cdot \sin\delta - \cos\delta \cdot \sin\phi \cdot \cos\tau)}{EH \cdot \cos\delta \cdot \sin\tau}$$

$$= \frac{\cos\phi \cdot \sin\delta - \cos\delta \cdot \sin\phi \cdot \cos\tau}{\cos\delta \cdot \sin\tau} .$$

The nomographic procedure which Ptolemy follows allows him to read the values of his coordinates directly from the analemma diagram using a straightedge in the form of a carpenter's square. Ptolemy divides the analemma into fixed and variable parts for a given latitude and designs an instrument which makes efficient use of this division in providing a diagram to which the straightedge can be applied. In his article on the analemma ("Das Analemma von Ptolemaeus," *Astronomische Nachrichten*, no. 230, 1927, pp. 17-46), P. Luckey gives a careful description of the instrument and explains why it can be considered a nomogram.

Nowhere in the preserved part of the ἀνάλημμα does Ptolemy explain why his six angles, computed for the seasonal hours, might be of interest to diallers. Commentators on the work have, however, suggested possible connections. In 1562, Frederico Commandino pointed out that horizon and descensivus angles determine shadow length and direction on a plane horizontal dial; that horarius and vertical angles determine shadow length and direction on a south (or north) facing vertical dial; and that hectemoros and meridian angles determine shadow length and direction on an east (or west) facing vertical dial. He presented detailed instructions for constructing dials with faces parallel to the orthogonal planes of Ptolemy's coordinate system in *De*

Horologiorum Descriptione published with *Liber de Analemmate* in 1562. Delambre (*Histoire de l'Astronomie Ancienne*, 1816) extended the comments of Commandino by suggesting that Ptolemy may also have considered angles which aided in the construction of vertical plane dials which deviate from the meridian and prime vertical planes.

A Catalogue of Greek and Roman Sundials

The catalogue contains all those stone sundials known to me which are thought to date from Greek and Roman antiquity through the fourth century A.D. Each included example marks the ancient seasonal hours. The descriptive material contained in each catalogue entry was obtained from published sources and, in a majority of cases, from personal examination of the dial. (The catalogue numbers of personally examined dials end in G.) The measurements that I made myself or selected from published materials are those which lead directly to a determination and analysis of the parameters of construction of the dial. My own measurements were made to the nearest millimeter; angular measurements were made to the nearest degree.

The format of each catalogue entry represents an attempt to maintain consistent presentation of details and to distinguish between observation and analysis. Measurements appear on the right in each entry; descriptive information and computed parameters appear on the left.

Bibliographic Abbreviations Used in the Catalogue

AEM--*Archäologisch-epigraphische Mittheilungen aus Österreich-Ungarn.*

AJA--*American Journal of Archaeology.*

Amelung--Amelung, Walther. *Die Sculpturen des Vaticanischen Museums.* Berlin, 1903-08.

Antonini--Antonini, Carlo. *Manuale di Varii Ornamenti,* vol. 4. Rome, 1781-90.

Ardaillon--Ardaillon, E. "Horologium." In *Dictionnaire des Antiquités Grecques et Romaines,* vol. 3, edited by C. Daremberg and E. Saglio, pp. 256-64. Paris, 1877-1919.

Ath. Mitt.--*Mittheilungen des deutschen archäologischen Instituts, Athenische Abteilung.*

AVN--*Annalen des Vereins fur Nassauische Alterthumskunde und Geschichtsforschung.*

AZ--*Archäologische Zeitung.*

BCH--*Bulletin de Correspondance Hellénique.*

Bracchi--Bracchi, L. C. "Orologi Solari di Aquileia." *Aquileia Nostra* 30(1960):50-69.

CIA--*Corpus Inscriptionum Atticarum.*

CIG--*Corpus Inscriptionum Graecarum.*

CIL--*Corpus Inscriptionum Latinarum.*

CIS--*Corpus Inscriptionum Semiticarum.*

Deonna--Deonna, W. *Le Mobilier Délien.* Exploration Archéologique de Delos, vol. 18. Paris, 1938.

Diels--Diels, Hermann. *Antike Technik.* Leipzig, 1920.

Drecker--Drecker, J. *Die Theorie der Sonnenuhren.* Berlin, 1925.

Gatty--Gatty, Mrs. Alfred. *The Book of Sundials.* London, 1900.

IG--*Inscriptiones Graecae.*

Ionescu-Carligel [1]--Ionescu-Carligel, C. "Cadrane Solare Grecesti si Romane in Dobrogea." *Pontice* 2(1969):199-208.

Ionescu-Carligel [2]--Ionescu-Carligel, C. "Contributions a l'étude des Cadrans Solaires Antiques." *Dacia* 14(1970): 119-37.

JA--*Journal Asiatique.*

Kenner--Kenner, F. "Römische Sonnenuhren aus Aquileia." *Milleilungen der K.K. Central-commission zur Erforschung und Erhaltung der Kunst-und Historische Denkmale* 6(1880): 1-20.

Petrie [1]--Petrie, W. M. F. *Naukratis I.* London, 1886.

Petrie [2]--Petrie, W. M. F. *Roman Portraits and Memphis*. British School of Archeology in Egypt and Egyptian Research Account, vol. 20. London, 1911.
RA--*Revue Archéologique*.
Rayet--Rayet, G. "Cadrans Solaires Coniques." *Annales de Chimie et de Physique* 6(1875):52-86.
Rehm--Rehm, A. "Horologium." In *Real-Encyclopädie der Klassischen Altertumswissenschaft* edited by Pauly, pp. 2416-28.
SEG--*Supplementum Epigraphicum Graecum*.
Schlieben--Schlieben, Major A.D. "Römische Sonnenuhren in Wiesbaden und Cannstadt." *Annalen des Vereins fur Nassauische Alterthumskunde und Geschichtsforschung* 20(1888):316-33.
Smith--Smith, A. H. *A Catalogue of Sculpture in the British Museum*. London, 1904.

Spherical Dials

Measured Quantities

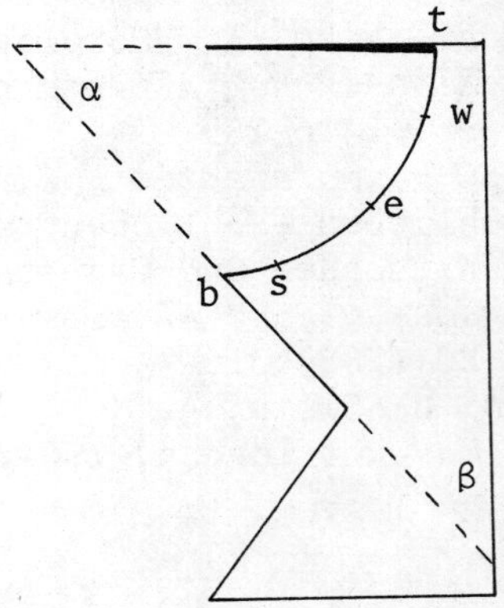

H = height (P = partial)
W = width
D = depth

tw = distance along meridian from top edge of dial to winter solstice curve
we = distance along meridian from winter solstice curve to equinox curve
es = distance along meridian from equinox curve to summer solstice curve
sb = distance along meridian from summer solstice curve to the bottom edge of the dial face

(VI to VII)
w = distance along winter solstice from the sixth to the seventh hour points
e = distance along the equinox from the sixth to the seventh hour points
s = distance along summer solstice curve from the sixth to the seventh hour points

α = acute angle between oblique front of dial and horizontal
β = acute angle between oblique front of dial and vertical

Calculated Quantities

r_t = radius of a trace of the top edge of the dial face
r_m = radius of the spherical dial face calculated from measurements taken along the meridian
r_e = radius of the spherical dial face calculated from measurements taken along the equinox curve

ϕ = intended latitude

1001. Spherical
Delos, Greece

H = 140 mm
W = 155 mm
D_{max} = 55 mm

Blue marble - Parts of three hour lines and three day curves are preserved on this fragment of spherical surface. The Greek inscription on the day curves has been restored by Dürrbach and Tannery as follows:

equinox:
[ἰση]μη[ρία]

above summer solstice:
ἐῴα ἐπι[τολή]

below summer solstice:
[τρ]οπαὶ θε[ριναί].

The destroyed winter solstice curve probably carried the following inscription:
[τροπαὶ χειμεριναί].

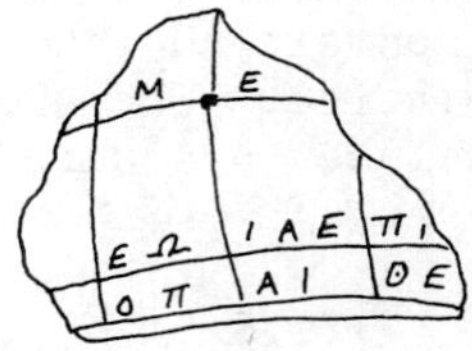

Found on July 9, 1903, in the Agora of the Delians on Delos, Greece. (Construction in the agora using blue marble began in the third century B.C.)

The arrangement of the plates in Deonna suggests that this fragment is on the island of Delos. I was unable to locate it in the spring of 1971.

Deonna, pp. 191-92, figure 228 (drawing) and plate 546 (photograph).
Dürrbach, F., and Jardé, A., "Fouilles de Délos," BCH 29(1905): 250-52 (drawing).
Tannery, P., *Mémoires Scientifiques*, vol. 14, Paris, 1937, pp. 277-83 (drawing).

1002G. Spherical
Delos, Greece
Museum, portico storage room

White tufa-like material covered with plaster - A break at the center of the dial has destroyed all of the meridian and parts of the fifth and seventh hour lines. All of the other hour lines and three day curves remain incised in the plaster coating. Traces of red color can be seen on the winter solstice line. Some hour lines extend into the area above this solstice. The simple overall design includes a squared back and sloping base.

H = 251 mm
W = 330 mm
D_{max} = 125 mm
D_{base} = 102 mm

(III to IV)
w = 18 mm
e = 32.5 mm
s = 45.5 mm

β = 39.5°

r_t = 137.5 mm

Deonna, p. 191, plate 541 (photograph).

***

1003. Spherical
Olympia, Greece
Archeological Museum, storeyard

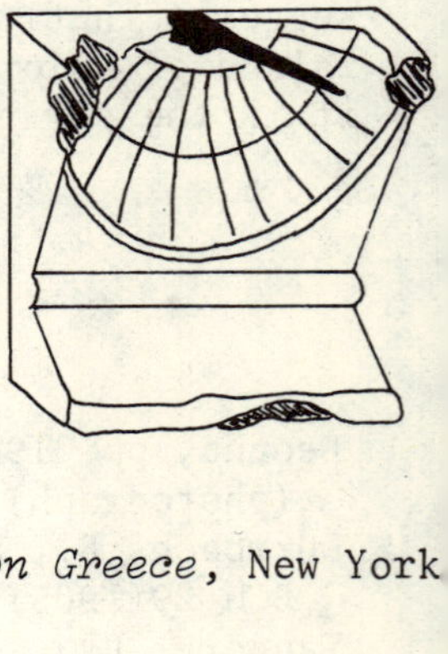

Complete bronze gnomon in position in the top surface of the dial.

White marble - Ten hour lines and three engraved day curves are visible in the photograph. A half-cylinder molding separates the oblique front from the base.

Moore, R. and Kazantzakis, N., *Reflections on Greece*, New York 1971 (photograph).

1004. Spherical
Kavalla, Greece
Museum

H = 425 mm
W = 287 mm
D = 210 mm
tc_1 = 60 mm
c_1c_2 = 22 mm
c_2c_3 = 25 mm
c_3c_4 = 23 mm
c_4c_5 = 25 mm
c_5b = 35 mm
(VI to VII)
c_1 = 12 mm
c_2 = 16 mm
c_3 = 20 mm
c_4 = 25 mm
c_5 = 29 mm
b = 35 mm

Brownish creamy stone - Eleven engraved hour lines divide six engraved day curves including one which coincides with the lower edge of the spherical surface. The dial is set on a very ornate base decorated with two lions' feet, and much Greek inscription.

Around dial face:
[Πάσας ὥ]ρας Δ̣ιο[γ]έν[ης ταῖς]
On sloping front:
Ἐννέα Μούσαις
Under slope:
[τ]ό[νδ]ε̣ τ̣ὸν ἡμέριον, Νεικομή
δης, κ̣ά̣μ̣νες Ἀρήτῃ,
Between legs:
[φ]ιλῶν, φίλε, ἐ
βίσκοπον
εὐγνώμονά
μ' ἠοῦς
εἵσατο Διο
γένης
οἷσι δόμοις
κτέ[α]ν̣[ον]

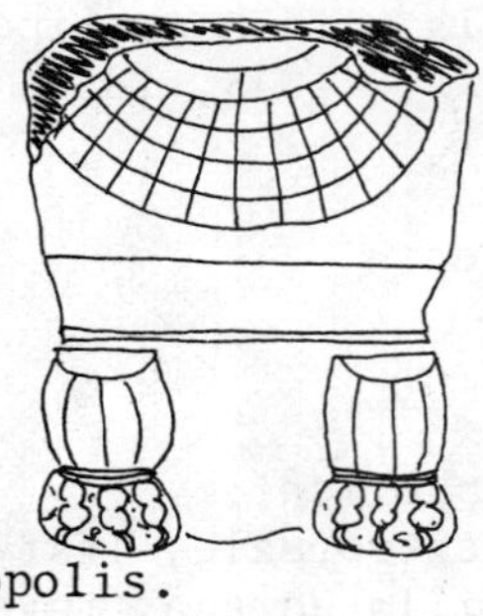

Found at Loutra, ancient Traianopolis.

Other spherical dials with lion feet include 1028G, 1056G, 1057G.

Measurements taken from Daphne Hereward's tracing.

Hereward, Daphne, "Inscriptions from Thrace," AJA 67(1963): 71-75 and plates 17-20 with drawings (SEG, XXIV, 642).
Peek, W., "Inschrift einer Sonnenuhr aus Traianopolis," *Zeitschrift für Papyrologie und Epigraphik* 10(1973): 98-100.

1005G. Spherical
Catania, Italy
Museo Civico

The gnomon hole is preserved in the top surface of the dial.

H = 325 mm
W = 321 mm
D_{top} = 220 mm

Limestone - The dial is cracked from the top to the bottom near the meridian. All eleven hour lines continue from the gnomon to the lower edge of the shadow-receiving surface. No day curves are visible. A half-cylinder molding separates the spherical surface from the sloping base.

tb = 165 mm

(VI to VII)
b = 48 mm

r_t = 137 mm

$\alpha = 47.2°$
$\beta = 41.2°$

Other dials with hour lines only include: 1006G, 1021G, 1046, 1052, 1063G.

Unpublished

***

1006G. Spherical
Castroreale, Italy
Villa Romana, inventory no. 71

A gnomon hole (20 mm x 50 mm) in the top surface contains no metal.

H = 315 mm
W = 415 mm
D_{top} = 246 mm
D_{base} = 157 mm

Pietra dolce de Siracusa - Eleven hour lines extend from the gnomon hole to the lower edge of the spherical surface. No day curves are visible.

tb = 197 mm

r_e = 158 mm

(VI to VII)
b = 58 mm

$\alpha = 51°$
$\beta = 38.2°$

Unpublished

1007G. Spherical
Aquileia, Italy
Museum, study collection

Plate 2
Page 128

A rectangular gnomon hole is visible in the top surface of the dial.

Porous stone - Eleven hour lines and an equinox are preserved. The summer solstice may coincide with the lower edge; the winter solstice may coincide with the boundary of the broken area at the top.

H = 230 mm
W = 302 mm
D = 130 mm

we = 34 mm
eb = 62.5 mm

(IV to V)
e = 22 mm
b = 34 mm

Unpublished

***

1008G. Spherical
Vienna, Austria
Kunsthistorisches Museum, no. I225

A gnomon hole in the top surface is 29 mm wide and 26 mm deep.

Limestone - Eleven hour lines extend between the winter solstice and the lower edge of the shadow-receiving surface. Two engraved day curves are visible.

From Aquileia.

r_t = 148 mm

H = 368 mm
W = 428 mm
D_{top} = 433 mm
tw = 45 mm
we = 105.5 mm
es = 104 mm
(VI to VII)
w = 22 mm
e = 40 mm
s = 49 mm
$\alpha = 60°$
$\beta = 30°$

Arneth, Joseph, *Beschreibung der zum K.K. Münz-und-Antiken-Cabinete gehörigen Statuen, Busten, Reliefs, etc.*, Vienna, 1866, p. 47, no. 216.
Cf. Bracchi, col. 60f, fig. 6 (photograph)
Kenner, p. 2, fig. 1 (drawing).

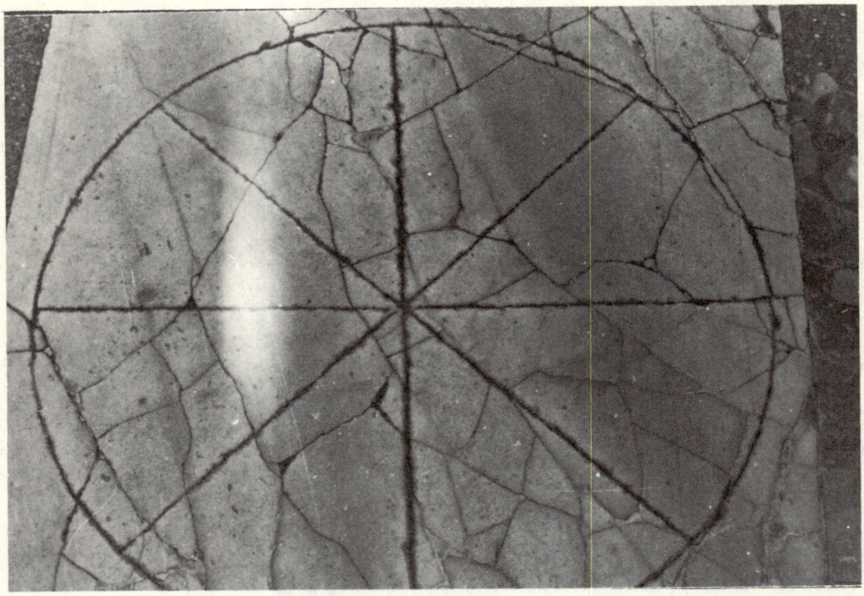
1

2

1009 Spherical
Present location unknown

Parian marble - Eleven engraved hour lines extend from the winter solstice to the summer solstice and are crossed by the equinoctial day curve. A half-cylinder molding separates the dial face from the base. Antonini's drawing shows a preserved gnomon.

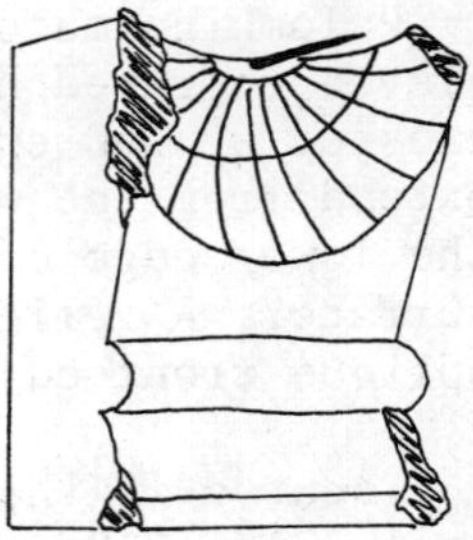

Seen in the Museo Borgiano in 1798.

Antonini, no. 7.

***

1010. Spherical
Present location unknown

White marble - The top surface of the dial is damaged so that only nine hour lines are preserved. Parts of three engraved day curves (equidistant from each other and the lower edge) are visible. The base is decorated with a tapering, knobbed, horizontal molding.

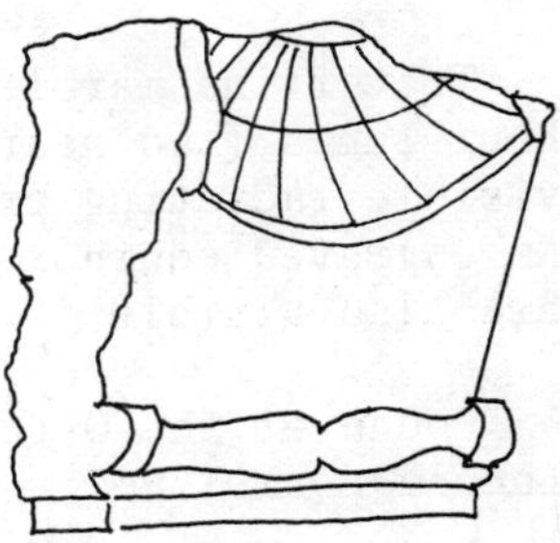

Seen in the Villa Borgia, Velletri, Italy, in 1798.

Antonini, no. 8.

1011. Spherical
Present location unknown

Palombino marble - Note that the eleven engraved hour lines are not crossed by an engraved equinox, but extend from the winter solstice to the lower edge of the shadow-receiving surface. A relief margin outlines the oblique front edge.

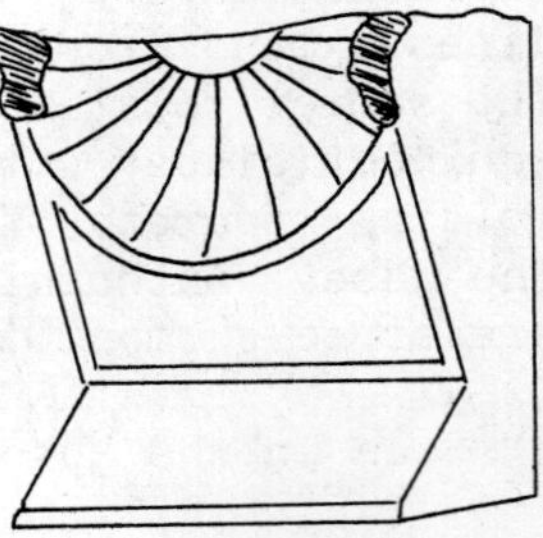

Located in the Casa Deangelis, Tivoli, Italy, in 1798.

Antonini, no. 9.

***

1012. Spherical
Present location unknown

Travertine marble - Thirteen engraved hour lines (two near the horizon) are visible radiating from a gnomon hole. An engraved equinox and summer solstice are also visible.

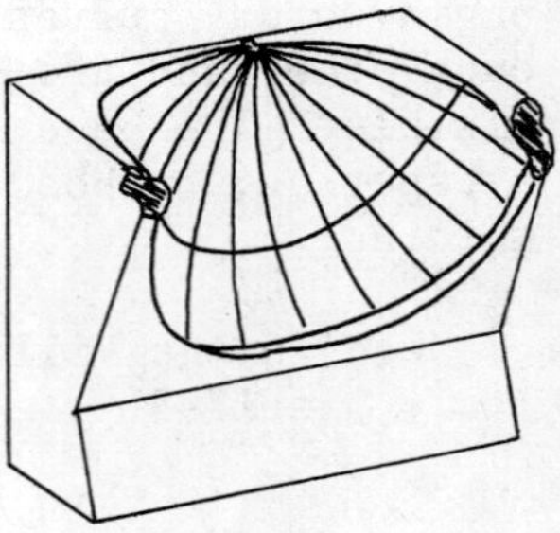

Found at the Osteria di Ponte Lucano near Tivoli.

In 1798, part of the collection of Thomas F. Hill, England.

Antonini, no. 16.

1013. Spherical
Present location unknown

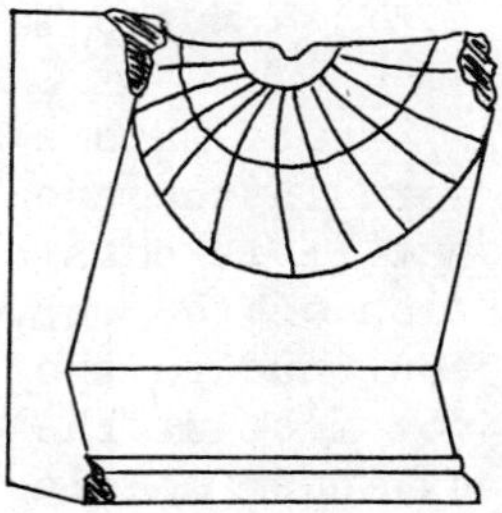

Travertine marble - Eleven engraved hour lines extend from the engraved winter solstice across the engraved equinox to the summer solstice which coincides with the lower edge.

Seen at the Casa Beschi, Tivoli, Italy, in 1798.

Antononi, no. 11.

***

1014. Spherical
Present location unknown

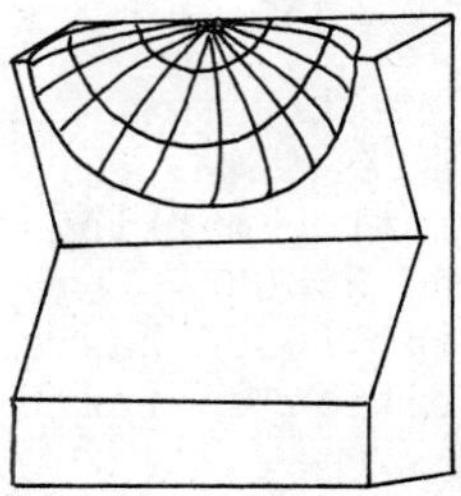

Eleven engraved hour lines radiate from the gnomon hole across the winter solstice and the equinox to the lower edge of the dial face. The summer solstice curve coincides with this lower edge.

Seen in Terracina (ancient Tarracina), Italy, in 1798.

Antonini, no. 13.

***

1015. Spherical
Larino, Italy
Cathedral

Found in the Villa Equitis de Blasus Aloysii and placed in the cathedral in 1891.

German Archeological Institute, Rome, negative no. 2194.

1016G. Spherical
Rome, Italy
Museo Nazionale Romano, study collection, no. 125596

White marble - The surface is badly worn in the area of the winter solstice, but it is possible to estimate the position of the curve. Eleven hour lines continue to the top of the dial but do not meet at the gnomon hole. A half-cylinder molding runs the width of the dial just above the rectangular relief blocks in the usual leg positions. The third, sixth, and ninth hour lines bear markings as shown in the diagram at right.

H = 490 mm
W = 465 mm
D = 355 mm
tw = 58.5 mm
we = 63 mm
es = 125 mm
sb = 5 mm
(VI to VII)
e = 44.5 mm
s = 78.5 mm
$\beta = 40°$

Found June, 1949, in Riano, Italy.

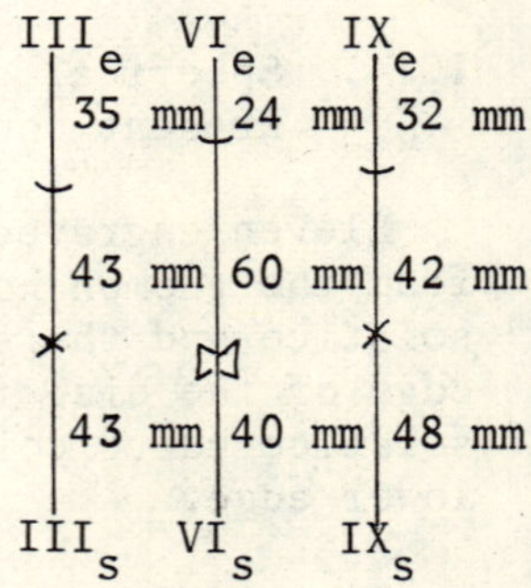

$r_t = 205$ mm
$r_e = 170$ mm

Similar markings are visible on 3067G and 3080G.
Dial 1036G has similar rectangular decoration on the base.

Unpublished

1017G. Spherical
London, England
British Museum, no. 2547

A gnomon hole in the top surface is 30 mm deep and has a horizontal cross section 30 mm x 17 mm.

Stone - All eleven hour lines continue from the gnomon hole to the summer solstice. Three engraved day curves are visible. The engraved summer solstice coincides with the lower edge of the spherical surface.

Found at Città Lavinia in 1891.

r_t = 118 mm

r_e = 97 mm

The winter solstice curve from hour 0 to hour XII is divided into the following arcs: 9, 11, 13.5, 13, 13.5, 13, 12.8, 13.5, 13.4, 13, 13.5, 8.5 mm. The equinox curve in divided into the following arcs: 21, 27.7, 28.2, 26, 24, 24, 24, 25, 27, 28.2, 27.7, 21 mm.

H = 235 mm
W = 278 mm
D_{top} = 184 mm
D_{base} = 112 mm

tw = 34 mm
we = 39 mm
es = 65 mm
sb = 4 mm

(VI to VII)
w = 12.8 mm
e = 24 mm
s = 42.2 mm

α = 55°
β = 42°

Cf. Gatty, p. 31.
Smith, vol. 3, p. 414, no. 2547.
Ward, F. A. B., *Handbook of the Collections Illustrating Time Measurement, Historical Review*, London 1947, p. 17, plate 2 (photograph) and *Descriptive Catalogue* London, 1950, p. 11.

1018. Spherical
Ince Blundell Hall
Lancashire, England

H = 220 mm

Luna marble - Eleven hour lines divide three day curves. A bust of a bearded man decorates the otherwise simple base. According to reports the bust was found near the excavated dial and appeared to belong where it is now placed.

Found at Palestrina before 1787.

An Account of the Statues, Busts, Bas-relieves, Cinerary Urns, and other Ancient Marbles, and Paintings at Ince Collected by H. B., Liverpool, 1803, no. 336, *Engravings and Etchings*, vol. 1, 1809, pl. 71 (engraving).

Ashmole, Bernard, *A Catalogue of the Ancient Marbles at Ince Blundell Hall*, Oxford, 1929, p. 55, no. 128, pl. 28 (photograph).

Michaelis, Adolph, *Ancient Marbles in Great Britain*, translated by C. A. M. Fennel, Cambridge, 1882, p. 297, also pp. 39-41.

Cf. Gatty, pp. 33-34.
Guattani, *Monumenti inediti*, 1787, April, pl. 2.2 (drawing).

1019G. Spherical
Pompeii, Italy
Granario

Plate 3
Page 137

The gnomon hole in the top surface is 49.5 mm deep.

Tufa - Only light traces of hour lines and day curves can be seen on the shadow-receiving surface. However, deeply engraved construction lines are clearly visible in the top and front surfaces. One straight line marks the meridian on the oblique front. At the top left, two lines have been engraved parallel to the back edge. The first is 68 mm from this back edge and the second is 45 mm from the first. A single line 68 mm from the back edge and also parallel to it is preserved at the top left. The traces of day curves visible on the shadow-receiving surface seem to coincide with these engraved guide lines at the top edge.

r_t = 118 mm

Other dials with construction guide lines in the top surface include 1030G, 3025G, 3028G.

H = 238 mm
W = 283 mm
D_{top} = 187 mm
D_{base} = 116 mm

tb = 126.5 mm

α = 47.8°
β = 42.5°

Unpublished

1020G. Spherical Plate 4
Pompeii, Italy
Granario

A tapering gnomon groove, 34 mm wide at the top, 14 mm wide at the bottom, and 32 mm deep pierces the top surface.

Gray porous stone - Eleven hour lines extend between engraved solstice lines and are crossed by the equinox. The day curves are equidistant from each other and the lower edge. In the oblique front of the dial a single engraved line follows the curve of the circular section 20 mm from the lower edge. A small round hole is visible in the right side.

Found along the Via di Nola, Region V, Insula III, no. 6.

r_t = 149 mm
r_e = 134 mm

H = 323 mm
W = 403 mm
D_{top} = 238 mm
D_{base} = 148 mm
tw = 42.5 mm
we = 61 mm
es = 87 mm
sb = 5 mm
(VI to VII)
w = 18 mm
e = 35 mm
s = 57 mm
$\alpha = 47.8°$
$\beta = 40°$

Unpublished

***

1021G. Spherical
Pompeii, Italy
Granario

Tufa covered with stucco - The top and sides are broken. Parts of eight hour lines are visible but no engraved day curves are preserved. Two schematized legs with cyma profiles flank the base.

r_t = 208 mm

PH = 460 mm
W = 379 mm
D_{base} = 219 mm

(V to VI)
b = 56 mm

Unpublished

3

4

1022G. Spherical
Pompeii, Italy
Museum, no. 2020/L

A rectangular gnomon hole, 24 mm long and 25 mm wide, lies in the narrow center of the top surface close to the top front edge.

Pentelic marble - The third, sixth and ninth of eleven hour lines are emphasized in a careless manner with red paint. Two engraved solstices limit the extent of the hour lines which are crossed by an equinox engraved always equidistant from both solstices and the lower edge. Two rosettes decorate the oblique front. A half-cylinder molding separates base from the spherical surface.

r_t = 144 mm
r_e = 157 mm

H = 341 mm
W = 335 mm
D_{top} = 256 mm
D_{base} = 118 mm

tw = 54 mm
we = 61 mm
es = 84.5 mm
sb = 4 mm

(VI to VII)
w = 21 mm
e = 41.2 mm
s = 55 mm

$\alpha = 51.8°$
$\beta = 37.5°$

Other spherical dials with half cylinder molding include 1003, 1005G, 1009, 1016G, 1024G, 1028G, 1059.

Mauri, Amadeo, *Pompeii*, Rome, 1970, p. 108.

1023G. Spherical
Pompeii, Italy
Granario

A copper gnomon now in position at the top of the dial is a restoration. However, visible lead traces are probably original.

Tufa - Seven hour lines (1,2,3, 4,5,6,7) and part of three day curves are preserved. A trace of red color can be seen at the junction of the meridian and the equinox. The dial was complete in 1960; now the right half is lost.

H = 260 mm
PW = 190 mm
W = 310 mm
PD = 170 mm

gw = 43 mm
we = 49.5 mm
es = 45 mm

(V to VI)
w = 15 mm
e = 27 mm
s = 38 mm

$\beta = 37.2°$

Other spherical dials with red color in the day curves include 1002G, 1024G, 1026G, 1027G.

Unpublished

1024G. Spherical
Pompeii, Italy
Granario

The base of a bronze gnomon remains in a lead filled gnomon hole in the top surface of the dial.

White marble with streaks of gray - Traces of red orange color remain in the eleven hour lines and three day curves. None of the hour lines extend beyond either solstice. A half-cylinder molding separates the shadow-receiving surface from the sloping base. Bronze and lead remain in the mount hole in the bottom surface.

r_t = 135 mm
r_e = 114 mm

H = 312 mm
W = 323.5 mm
D_{top} = 192.5 mm
D_{base} = 134 mm

tw = 44 mm
we = 40 mm
es = 84.8 mm
sb = 7.5 mm

(VI to VII)
w = 18 mm
e = 30 mm
s = 48 mm

α = 51.5°
β = 38.5°

For other spherical dials with half-cylinder molding see 1003, 1005G, 1009, 1016G, 1022G, 1028G, 1032G, 1059.

Unpublished

1025G. Spherical
Pompeii, Italy
Region II, Insula XI, no. 7

A rounded gnomon hole in the top surface is 25 mm deep.

Tufa covered with stucco - All eleven hour lines and the three day curves have been deeply engraved in the tufa. Traces of dark red color emphasize the hour lines which extend beyond both solstice curves.

r_t = 149 mm

r_e = 134 mm

H = 328 mm
W = 321 mm
D_{top} = 283 mm
D_{base} = 229 mm

tw = 40 mm
we = 46 mm
es = 68 mm
sb = 10 mm

(VI to VII)
w = 18.5 mm
e = 35 mm
s = 49 mm

$\alpha = 47°$

Unpublished

1026G. Spherical
Pompeii, Italy
Granario

Plate 5
Page 144

A gnomon hole 34 mm deep remains in the top surface of the dial.

Tufa covered with stucco - A deep red color marks the eleven hour lines and three day curves. The same red color fills in the area above the winter solstice; and traces of red are visible on the sloping base.

r_t = 177 mm
r_e = 164 mm

Red paint is used to decorate two other tufa and stucco dials found at Pompeii: 1029G and 1030G.

H = 303 mm
W = 439 mm
D_{top} = 274 mm
D_{base} = 172 mm

tw = 44 mm
we = 69.5 mm
es = 89.5 mm
sb = 4 mm

(VI to VII)
w = 23 mm
e = 43 mm
s = 56 mm

$\beta = 40°$

Unpublished

1027G. Spherical
Pompeii, Italy
Granario

Plate 6
Page 144

The original gnomon hole has been filled by a restorer, and a thin wire gnomon has been inserted.

Tufa - Eleven hour lines emphasized with dark red color extend from the gnomon hole to the lower edge of the shadow-receiving surface. The same dark red color can be seen in the engraved winter solstice and equinox. The base is broken away.

r_m = 133 mm, for ε = 24°

r_e = 95 mm

PH = 200 mm
W = 290 mm
D_{top} = 193 mm

tw = 46 mm
we = 56 mm
eb = 55 mm

(VI to VII)
w = 12 mm
e = 25 mm
b = 40 mm

α = 55°

Other spherical dials with red in hour lines include 1024G, 1025G, and 1026G.

Unpublished

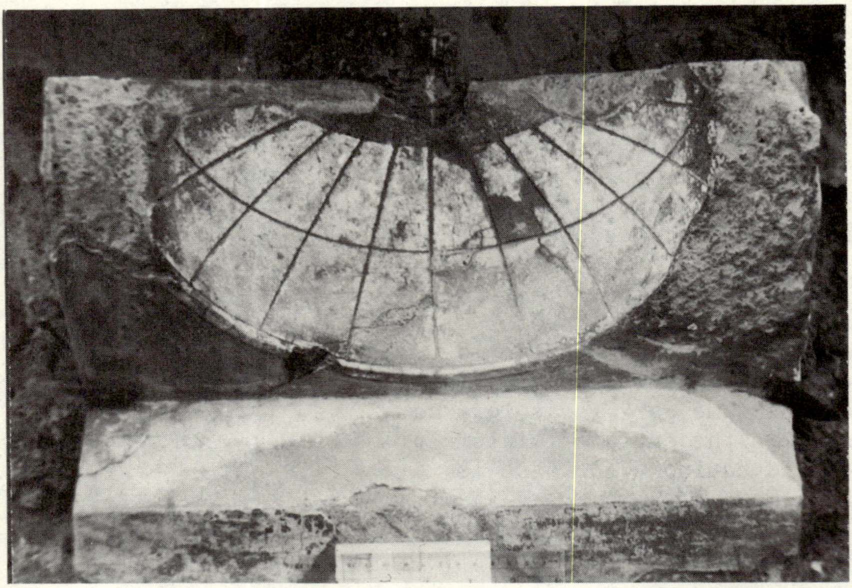

5

6

1028G. Spherical
Pompeii, Italy
Region I, Insula X, no. 4, House of the Menander

The base of a bronze gnomon in lead fill is preserved in the top of the dial surface.

White marble - Eleven hour lines continue from the gnomon groove to the lower edge. Two engraved day curves are visible. A half cylinder and two four-toed lion paws decorate the base; a clamp groove is preserved in the lower center front. Traces of iron and lead can be seen in a mount hole marking the center of the bottom surface.

The house dates from the period between 80 B.C. and 14 A.D.

r_t = 131 mm
r_e = 105 mm

Other spherical dials with cylindrical molding include 1003, 1005G, 1009, 1022G, 1024G, 1032G, 1059.

Unpublished

H = 423 mm
W = 337 mm
D = 215 mm

tw = 36.5 mm
we = 56 mm
eb = 83 mm

(VI to VII)
w = 13 mm
e = 27.5 mm
b = 43 mm

α = 59°
β = 34.5°

1029G. Spherical
Pompeii, Italy
Region III, House of the Count of Turin
(Domus M. Obelli Firmi)

Tufa covered with red painted stucco - No color is preserved on the dial face. Only seven hour lines (3,4,5,6,7,8,9) and an engraved equinox are visible. Top and sides are broken away. A shelflike molding extends from the left side.

The house dates from the period between 30 B.C. and 32 A.D.

PH = 480 mm
W = 650 mm
D_{base} = 270 mm
eb = 133 mm
(VI to VII)
e = 66 mm
b = 92 mm
$\beta = 43.5°$

Other dials with decorative red paint have been found at Pompeii: 1026G and 1030G.

Unpublished

1030G. Spherical
Pompeii, Italy
Granario

Plate 7
Page 165

Preserved portions of the gnomon hole indicate a depth of 28 mm.

Tufa covered with stucco - A trace of red coloring remains on the sloping base. A construction guide line engraved in the top surface is clearly visible parallel to the back edge at a distance of 97 mm. Only half of the shadow-receiving surface is preserved. On it, five hour lines (7,8,9,10,11) and two day curves (winter solstice and equinox) can be seen engraved in the stucco. The top guide line does not meet the winter solstice line, although the two lines intersect the top edge very near to each other.

r_t = 105 mm

Other Pompeian dials with red coloring include 1026G and 1029G.

Unpublished

H = 244 mm
½W = 135 mm
D_{top} = 225 mm
D_{base} = 130 mm
tw = 37 mm
we = 49 mm
eb = 54 mm

(VII to VIII)
w = 19 mm
e = 32 mm
b = 41.5 mm

α = 48°

1031G. Spherical
Herculaneum, Italy
Casa del Atrio a Mosaico

The gnomon in position is a restoration.

White marble - None of the eleven hour lines continue beyond either solstice. The equinox is always equidistant from the solstice curves and from the lower edge of the shadow-receiving surface.

r_t = 185 mm
r_e = 143 mm

H = 273.5 mm
W = 408 mm
D_{top} = 241 mm
D_{base} = 134 mm

tw = 41 mm
we = 68 mm
es = 83 mm
sb = 4 mm
(VI to VII)
w = 18 mm
e = 37.5 mm
s = 62.5 mm
$\alpha = 48.5°$
$\sigma = 83.2°$
$\gamma = 52°$
$\beta = 41°$

Rayet, p. 81, no. 407.

1032G. Spherical
Ostia Antica, Italy
Museum, no. 1447

The horizontal section of the gnomon hole is a narrow rectangle 25 mm x 6 mm.

Travertine marble - The eleven hour lines do not continue beyond either solstice. Three engraved day curves are visible. A half-cylinder molding separates the oblique front surface from the sloping base.

Found on the Decumanus Maximus near the baths of Neptune.

r_t = 107 mm
r_e = 96 mm

A half-cylinder molding also decorates 1003, 1005G, 1009, 1024G, 1028G, 1059.

H = 270 mm
W = 258 mm
D_{top} = 175 mm
D_{base} = 110 mm

tw = 35 mm
we = 34.5 mm
es = 67 mm
sb = 3 mm

(VI to VII)
w = 16 mm
e = 25.2 mm
s = 36.2 mm

$\alpha = 50°$
$\beta = 40°$

Calza, R. and Squarciapino, M. F., *Museo Ostiense*, Rome, 1966, p. 17.

1033G. Spherical
Ostia Antica, Italy
Magazine, no. 14463

The gnomon hole is 17 mm deep.

Cream marble - The top left corner is broken away. Nine hour lines are visible. Only the meridian continues beyond the winter solstice to the gnomon hole. Three day curves are preserved.

r_t = 110 mm
r_e = 103 mm

H = 239 mm
W = 258 mm
D_{top} = 210 mm
D_{base} = 102 mm
tw = 35 mm
we = 46 mm
es = 58 mm
sb = 6 mm
(VI to VII)
w = 14 mm
e = 27 mm
s = 38 mm
α = 48.5°
β = 41.2°

Unpublished

***

1034. Spherical
Present location unknown

Eleven engraved hour lines radiate from a small engraved circular arc near the gnomon hole. The lines cross three engraved parallel circles. The third, sixth, and ninth hour lines extend to the lower edge of the dial face. The spherical face is supported by a kneeling Atlas.

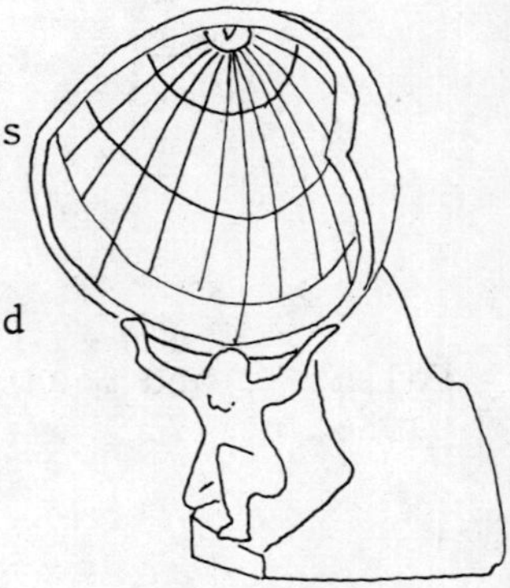

Found at Tor Paterno, near Ostia, in 1790.

Settele, G., *Memoria sopra la forma delle linee orarie indicanti le ore ineguali degli antichi sopra gli orologi solare*, Rome, 1816, p. 81 and frontispiece (drawing).
Cf. Gatty, p. 36.

1035G. Spherical
Rome, Italy
Museo Nazionale Romano, courtyard

A gnomon hole, 30 mm in diameter, is visible in the top surface of the dial.

White marble - The base of this dial is severely simple. Eleven hour lines cross three day curves. The length of the hour lines is limited by the solstice curves.

r_t = 182 mm

r_e = 126 mm

This dial should possibly be identified with that published by Luca Zuzzeri in 1746.

H = 334 mm
W = 427 mm
D = 240 mm

tw = 48.5 mm
we = 50 mm
es = 96.5 mm
sb = 6 mm

(VI to VII)
w = 17.5 mm
e = 33.8 mm
s = 59 mm

α = 50°
β = 38.8°

Cf. Portaluppi, Piero, *Gnomonica Atellana*, Milan, 1967, p. 19 (photograph).
Antonini, no. 4.
Zuzzeri, Luca, *Dissertazione d'un Antico Orologio a Sole*, Venice, 1746, p. 62 and plate (drawing).

1036G. Spherical
Rome, Italy
Palazzo dei Conservatori, upper courtyard

The green and rusty gnomon now in place is probably not original. However, beneath the ring base of the present gnomon there is a hole which may have held the original gnomon.

Travertine marble, white - Stylized legs in low relief flank the base. A 12 mm wide lip follows the edge of the front circular section. Eleven hour lines divide three day curves.

Found near Castelnuovo, Italy on the Via Flaminia in 1751. Pope Benedict XIV restored the gnomon and fixed it in position.

r_e = 172 mm

Dial 1016G has similar rectangular stylized legs.

H = 380 mm
W = 435 mm
D_{top} = 267 mm
D_{base} = 175 mm
gw = 47 mm
we = 66 mm
es = 122 mm
sb = 9 mm
(VI to VII)
w = 25 mm
e = 45 mm
s = 70 mm

$\alpha = 55°$
$\beta = 34°$

A Catalogue of the Ancient Sculptures Preserved in the Municipal Collections of Rome, The Sculptures of the Palazzo dei Conservatori, Oxford, 1926, p. 238.
Nuova Descrizione del Museo Capitolino, Rome, 1882.
Cf. Antonini, dial no. 1.
Drecker, p. 21, figure 27 (photograph).

1037G. Spherical
Vatican City
Vatican Museum, Cortile Ottagono, no. 939

A damaged gnomon hole is visible at the top above the meridian.

White marble - Of all eleven hour lines, only the meridian continues to the gnomon hole. The three day curves are parallel to each other. A horizontal crescent has been engraved on the oblique front surface below the meridian, and two six leafed rosettes flank it.

r_m = 131 mm
r_e = 122 mm

H = 180 mm
W = 297 mm
D = 138 mm

gw = 21+ mm
we = 55 mm
es = 55 mm

(VI to VII)
w = 19 mm
e = 32 mm
s = 45 mm

α = 51°

Rosette and crescent design also on 1059, 3022, 3067G.

Amelung, p. 106.

1038G. Spherical
Vatican City
Vatican Museum, Cortile Ottagono

A rectangular gnomon hole is preserved in the center of the top surface of the dial.

Travertine marble - Three parallel day curves and eleven hour lines form the usual grid. All hour lines continue beyond the winter solstice to the gnomon.

H = 222 mm
W = 255 mm
D_{top} = 167 mm
gw = 25 mm
we = 40 mm
es = 61 mm
sb = 8 mm
(VI to VII)
w = 10 mm
e = 28 mm
s = 37 mm
α = 49°

Unpublished

***

1039. Spherical
Volubilis, Morocco
Musée Lapidaire en Plein Air

Light color stone - Although both sides are broken away, parts of eleven hour lines remain to divide three day curves.

H = 460 mm
W = 560 mm
D_{top} = 160 mm

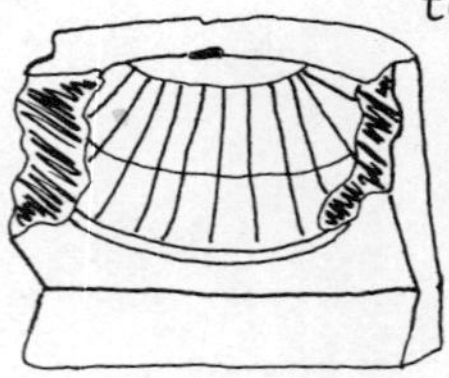

Measurements and a brief description from L. Janin.

1040G. Spherical
London, England
British Museum, no. 1909 2-16 10

A gnomon groove 5 mm wide remains in the top surface of the dial.

Limestone - This miniature dial has a rounded back. Twelve hour lines extend from a day curve near the gnomon to the lower edge of the spherical surface. Traces of engraved letters are visible below the spherical surface.

Found at Naukratis in 1883.

r_m = 33.8 mm

Other dials with only two day curves include 1011, 1041, 1059G, 1067G.

The winter solstice curve is made up of arcs of the following lengths:
2, 2.5, 3.7, 4.5, 4.9, 4, 2.9, 3.1, 4, 3.9, 2.5, 3.9, 2 mm.
The curve at the lower edge of the dial is made up of arcs of the following lengths:
2, 9.5, 9.5, 11.5, 10.5, 11, 8.9, 10, 8, 8.5, 9, 9, 7.5 mm.

H = 65 mm
W = 78 mm
D_{top} = 50 mm
D_{base} = 34 mm

tw = 12.5 mm
wb = 37 mm

(VI to VII)
w = 2.9 mm
b = 8.9 mm

α = 83.7°

Petrie[1], p. 16 and plate 18.6 (drawing).

1041. Spherical
Tripoli, Libya

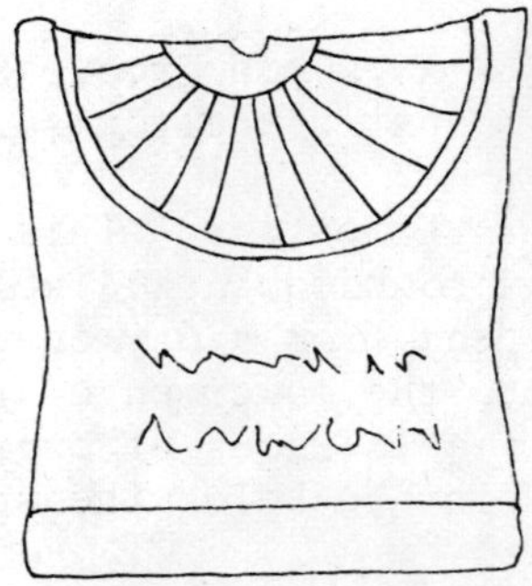

Eleven hour lines extending between two day curves are visible in the sketch by L. Janin reproduced at right. A neo-Punic inscription is preserved in two lines below the dial face.

אידת ארשם בן בעלשלך

ת
הבנא כתצאתם בתם תהפ סת
ז

Found at Leptis Magna, Libya.

Dated from epigraphy to the first century A.D.

For other spherical dials with divided solstices only see 1011, 1042, 1067G.

della Vida, G. L., "Un Inedita Inscriptione Neopunica da Leptis Magna (Tripol 52)," *Wissenschaftliche Zeitschrift Univ. Halle* 17 (1968): 127-32 (photograph).

1042. Spherical
Tunis, Tunisia
Musée National du Bardo

White stone - The meridian line extends from the gnomon hole to the bottom of the base. Other preserved hour lines extend from the winter to the summer solstice. There is no engraved equinox curve.

Drecker, p. 55, fig. 84 (photograph)

***

1043. Spherical
Merida, Spain

Half of a limestone dial was found in the ruins of the Roman theater at Merida. Five hour lines and three day curves are visible in the published photograph.

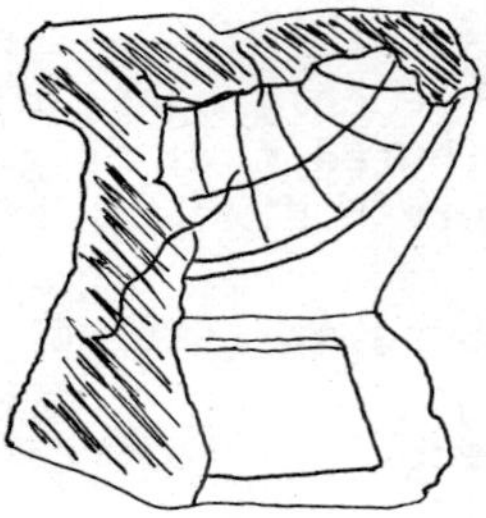

Melida, J., "El Teatro Romano de Merida," *Revista de Archivos, Bibliotecas y Museos* 3 (1915): 36, plate 5 (photograph).

1044. Spherical
Bucharest, Rumania
National Museum of Antiquities, no. L2023

H = 121.5 mm
r = 157 mm

es = 66.5 mm

(V to VI)
e = 41.2 mm
s = 48.5 mm

White marble - Only a fragment of the shadow-receiving surface of this dial is preserved; but on it, two hour lines, including meridian, are visible. The equinox and the summer solstice day curves can also be seen as well as part of a triangle having vertex at the winter solstice curve and the meridian as axis. Part of a Greek inscription is preserved along the triangle line: ος] χρό[νος] Thirteen short lines marked with the Greek letters [α] through [ν] divide the meridian between winter and summer solstices. On the basis of the visible epigraphy, E. Popescu has attributed this fragment to the third century B.C.

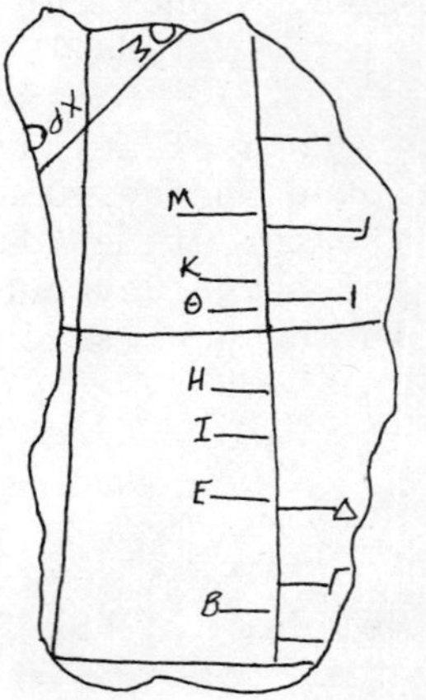

Found at Istropolis on the Black Sea.

Ionescu-Carligel calculates ϕ to be 44°31'.

For $\varepsilon = 24°$

r_m = 159 mm

r_e = 157 mm

Other dials marked with a daylight triangle include 1068G, 3046, 4001G. See 4001G for completion of partially preserved inscription.

Ionescu-Carligel [2], pp. 121-25, fig. 2 (photograph), pp. 134-37, figs. 10, 11 (drawings).
Ionescu-Carligel [1], pp. 199-201, fig. 1 (photograph), fig. 2 (drawing), pp. 206, 207, fig. 5 (drawing).

1045. Spherical
Nîmes, France
Musée Archéologique, no. 908-15-1

Traces of a gnomon hole are visible in the top surface of the dial above the meridian.

H = 400 mm
W = 330 mm
D = 270 mm

White limestone - Although both sides are broken, eleven engraved hour lines and two engraved day curves are preserved. The summer solstice coincides with the lower edge of the spherical surface. The large base is quite plain.

Found in the ruins of a Roman villa at Renisulius in 1908.

Mémoires de l'Académie de Nîmes, 1908, pp. 267-68.

***

1046. Spherical
Grenoble, France
Musée Dauphinois

Reused in the building of the Vienna gate of ancient Gratianopolis Cularo in the third century A.D.

H = 800 mm
W = 560 mm
D = 560 mm
r = 280 mm

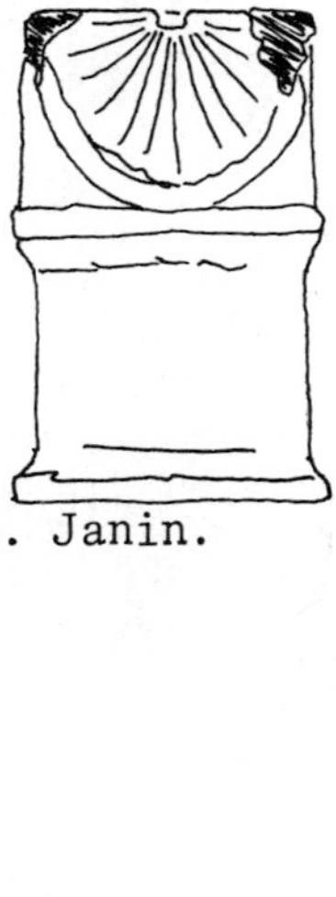

Measurements and sketch from L. Janin.

1047. Spherical
Nissan-les-Euserune, France
Private collection of Father J. Giry, curator of the National Museum of Euserune

A well-preserved bronze gnomon is in position in the top surface of the dial above the meridian.

H = 300 mm
W = 210 mm
D = 250 mm

Eleven well-traced hour lines and three day curves divide the shadow receiving surface.

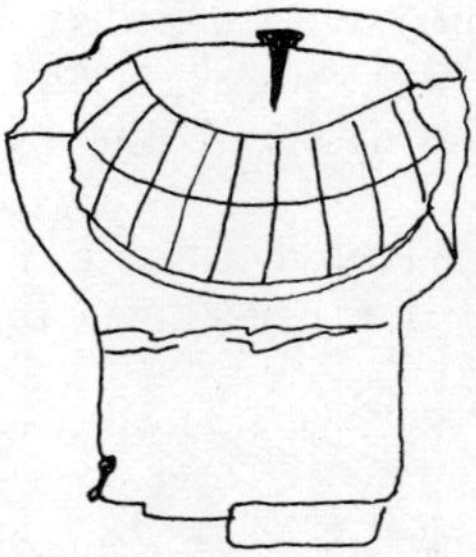

Found at Quarante among the ruins of Roman villas dating from the third or fourth century A.D.

Janin measures a latitude of 37° with obliquity of ecliptic equal to 24°.

Measurements and sketch sent to me by L. Janin, Sèvres, France, on February 9, 1971.

1048G. Spherical
Bucharest, Rumania
National Archeological Museum,
no. MNA L990

A gnomon hole (27 mm long x 15 mm wide x 49 mm deep) is preserved in the back of the dial with a 20 mm wide groove in the top surface.

Limestone - Eleven hour lines extend from the winter solstice across the equinox to the lower edge. A twelfth lightly scratched "hour" line runs parallel to the meridian 4.5 mm to the right. Two hands and a damaged head support the rounded sides, back and bottom of the dial surface.

H = 225 mm
W = 2 x 140 mm
W_{max} = 290 mm
D = 168 mm
r = 128 mm
tw = 72 mm
we = 59 mm
eb = 57.5 mm
(V to VI)
w = 16 mm
e = 28.1 mm
b = 36.2 mm
$\alpha = 52.8°$

G. Bordenache dates workmanship to the third century A.D.
Ionescu-Carligel calculates the latitude of the dial to be 44°15'.

Ionescu-Carligel [2], (photograph).

1049. Spherical
Strasbourg, France
Archeology Museum

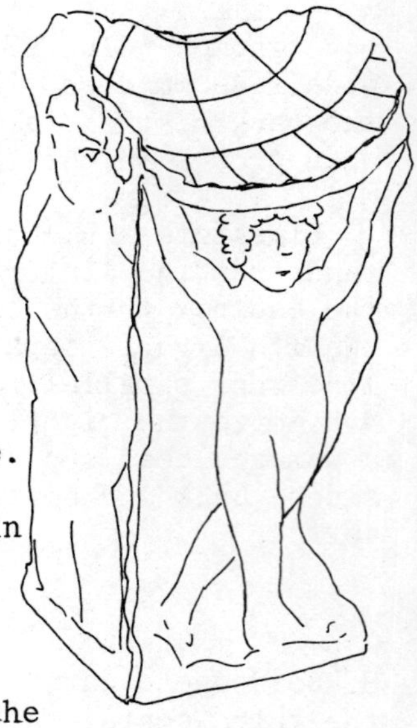

Streaked sandstone from the Vosges Mountains - Three engraved day curves are visible. Possibly a fourth day curve coincided with the lower edge of the spherical shadow-receiving surface. The meridian and two other hour lines extend from the first day curve to the lower edge. Between the third day curve and the lower edge, the areas between major hour lines are each divided into three parts, so that twelve hours are marked in all. Four standing figures decorate the base.

Found at the village of Bettwiller in the spring of 1879.

The style of the sculpture dates from the second century A.D.

Rohr, René, "Le Cadran Solaire Romain de Bettwiller," *Cahiers Alsaciens d'Archéologie d'Art et d'Histoire*, 1966, pp. 50-54 (drawing).

1050G. Spherical
London, England
British Museum, no. 2548

The gnomon hole is 29 mm deep, and is emphasized by a relief margin at back.

H = 184 mm
W = 214 mm

r_t = 111 mm

Stone - Eleven hour lines radiate from the gnomon hole to the lower edge of the spherical surface. One engraved day curve is visible always equidistant from the lower edge. The sides and back are rounded.

tc = 88.8 mm
cb = 79 mm

(VI to VII)
c = 16.5 mm
b = 27.5 mm

For r = 111 mm, $\overline{\phi}$ at c = 48°.

α = 93.5°

The curve c from horizon to horizon is made up of the following arc lengths: 18, 19.2, 18.5, 16, 17.5, 16.5, 16.5, 17, 18.5, 19.5, 22,21 mm.
The lower edge of the dial face from the third to the ninth hour line is made up of the following arc lengths: 27.5, 27, 29, 27.5, 25, 27.5 mm.

Smith, vol. 3, p. 414, no. 2548.
Cf. Cousins, F. W., *Sundials*, New York: Pica Press, 1970, p. 25 (photograph).

1051G. Spherical Plate 8
Istanbul, Turkey
Archeological Museum, no. 4845

A gnomon hole with lead is preserved in top surface of the dial above the meridian.

White marble - This miniature dial is part of a funeral stele of the philosopher Theodotos, son of Menephron, dating from the second century B.C. Ten hour lines are preserved, converging toward the gnomon groove. Only one engraved day curve crosses these hour lines. A tiny sphinx supports the shadow-receiving surface. A similar sphinx in the upper right corner probably also supported a sundial.

dial
H = 170 mm
W = 2 x 62 mm
D = 47 mm

tc = 49 mm
cb = 28 mm

(VI to VII)
c = 12 mm
b = 18.5 mm

The dial face is a quarter sphere.

For r = 49 mm, c corresponds to $\overline{\phi} = 57°29'$.

See 1050G for another quarter spherical dial with only one day curve.

Firatli, Nezih, "Les Stèles Funeraires de Byzance Gréco-Romaine," *Bibliothèque Archéologique et Historique de l'Institut Français d'Archéologie d'Istanbul*, 1964, p. 54, pl. 8 (photograph).

7

8

1052. Spherical
Carthage, Tunisia
Musée de Carthage

Stone - The shadow-receiving surface consists of two separate half quarter spheres joined to make a complete quarter sphere. The gnomon point is the corner of the solid juncture surface. Ten hour lines (five in each segment) stretch from the top center of the spherical surface to the lower edge. No day curves are visible.

H = 350 mm
W = 310 mm
D = 155 mm
r = 115 mm

Other spherical dials without day curves include 1005G, 1006G, 1021G, 1046, 1063G.

Measurements and photograph from Ennaifer Mongi, Curator of the Carthage museum.

1053G. Spherical
Thessalonica, Greece
Archeological Museum, study collection

A wedge-shaped hole at the back and a groove in the top surface once held the gnomon.

Porous gray stone - Eleven hour lines extend from the gnomon hole to a semicircle which has been engraved some distance from and not parallel to the lower edge of the dial. A framed area has been prepared, as if for inscription, on the base.

H = 393 mm
W = 243 mm
D = 163 mm

tc = 90 mm
cb = 215 mm

(VI to VII)
c = 21 mm

The radius of the engraved circle is 88 mm.

The curve c from the first hour line to the right horizon is made up of the following arc lengths: 21.5, 20, 20, 20, 21, 21, 19, 21, 21.5, 20 mm.

A similar, but much more ornate dial was found at Silistra, Bulgaria, (ancient Dorostorum). See catalogue entry 1054G.

Unpublished

1054G. Spherical
Silistra, Bulgaria
Archeological Museum, no. 517

Plate 9
Page 170

A rectangular gnomon hole is preserved in the front surface of the dial, along the meridian 20 mm from the top edge. Its dimensions are 27 mm long x 10 mm wide x 22 mm deep.

White limestone - A crude grid of hour lines and day curves is combined with an elaborate relief design to form this dial. In all, thirteen hour lines radiate from the gnomon hole and continue to intersect the thin crescent formed by curves c_1 and c_2. Hour lines three and nine continue to intersect a second crescent formed by curves c_3 and c_4. A wide lip emphasizes the circumference of the front section. The relief depicts a youthful figure playing a harp in a garden with various animals.

H = 880 mm
W = 830 mm
D = 296 mm

tg = 25 mm
gc_1 = 110 mm
c_1c_2 = 27 mm
c_2c_3 = 57 mm
c_3c_4 = 25 mm
c_4b = 104 mm

(V to VI)
c_1 = 30 mm
c_2 = 32.5 mm

r_{front} = 288 mm

Found in the ruins of ancient Dorostorum.

The dial face is a flattened quarter sphere.

r_m = 221 mm

Mladenova attributes the art work to the end of the second or the beginning of the third century A.D.

Note the similarity of this dial to the much simpler 1053G.

Mladenova, I., "Solnechnye Chasy S Rel'efom Orfeya Iz Durostoruma," *Sovetskaya Arkheologiya* 4(1961): 274 (photographs).

1055G. Spherical — Plate 10, Page 170
Selçuk, Turkey
Archeological Museum, courtyard, no. 375

The present gnomon is not original, and museum restoration has destroyed any trace of the original gnomon hole.

White marble with gray streaks - The eleven hour lines radiate from the gnomon hole across one engraved circle to a second. The day curves are parallel to each other but do not follow the curve of the lower edge. Marks in the dial surface above the two clearly visible day curves may form a third curve. (I have designated it c_1 in the measurements.) The spaces between hour lines are inscribed with large Greek letters:

A B Γ Δ E S Z H Θ I IA IB.

The long Greek dedication on the base was inscribed with a different hand:

Αὐτοκράτορι
Καίσαρι Μ. Αὐρ(ηλιῳ)
Σεουήρῳ
Ἀντωνείνῳ
Σεβ(αστῳ) καὶ Ἰουλίᾳ
[Σ]ε[β]αστῇ μητρὶ
[Κάστρων---------]

$H = 580$ mm
$W_{top} = 1180$ mm
$D_{top} = 450$ mm
$tc_1 = 164$ mm
$c_1c_2 = 142$ mm
$c_2c_3 = 165$ mm
$c_3b = 184$ mm

(VI to VII)
$c_1 = 54.2$ mm
$c_2 = 93.2$ mm
$c_3 = 132$ mm

$\alpha = 84°$

The dial is dedicated to the Emperor Caracalla (Marcus Aurelius Severus Antoninus) and his mother Julia Domna.

Gatty says the dial was found at Aphrodisias before 1877.

For $\alpha = 84°$, $r_m = 447$ mm.
tc_2 subtends an angle of 39° at the center of the sphere.
tc_3 subtends an angle of 60° at the center of the sphere.

Miltner, Franz, "Vorlaüfiger Bericht über die Ausgrabungen in Ephesos," *Jahreshefte des Österreichischen Archäologischen Institutes in Wien* 42(1955) : beiblatt columns 56-58, figure 20 (photograph).
Cf. Gatty, p.32.

9

10

1056G. Spherical
London, England
British Museum, no. 2545

Traces of iron and lead remain in the gnomon hole.

Marble - Of eleven hour lines, only the third, sixth, and ninth continue beyond the summer solstice to the lower edge of the shadow-receiving surface. Three engraved day curves, always equidistant from each other, are visible. The day curves are not always equidistant from the oblique front edge. A chord from the top of the dial to the lowest point of the meridian has a sagitta of 10 mm at the equinox. Traces of bored holes in the eyes and mouth can be seen in the lion heads on four-toed paws which flank the base.

H = 536 mm
W = 476 mm

tw = 113 mm
we = 82 mm
es = 81 mm
sb = 45 mm

(VI to VII)
w = 42 mm
e = 62 mm
s = 75 mm

$\alpha = 90°$
$\beta = 30°$

$r_m = 195$ mm

$r_e = 212$ mm, for $\varepsilon = 24°$.

The equinox curve is composed of arcs of the following lengths:
40.5, 58.5, 58, 51.5, 61.5, 59,
53, 55, 56, 54.5, 58, 62 mm.
The summer solstice curve is made up of arcs of the following lengths from the second to the tenth hour line:
73, 73, 77, 73, 72, 72, 72, 75 mm.

See 3041 for a dial with similar decorative details.

1057G. Spherical
Paris, France
Louvre Museum, no. MA 839

The gnomon hole is 28.5 mm deep.

White marble - The network of lines on this dial does not seem to be integrally related to the shape of the shadow-receiving surface. The surface itself is a flattened segment of a sphere, oblique and horizontal sections being circles of radius about 121 mm. Eleven hour lines radiate from the gnomon hole to the lower edge of this surface. Three engraved day curves cross these hour lines. The day curves are equidistant from the lower edge of the flattened segment. Two five-toed lion feet flank a base decorated with a central four-petalled, four-leafed flower.

H = 311 mm
W = 299 mm
D_{top} = 265 mm
D_{base} = 230 mm

tw = 53 mm
we = 32 mm
es = 33 mm

(VIII to IX)
w = 11 mm
e = 19 mm
s = 28 mm
b = 34 mm

$\alpha = 44°$

At one time in the Musée de Marseille.

For other spherical dials with lion feet, see 1004, 1028G, 1056G.

Millin, A. L., *Voyage dans les departemens du Midi de la France*, vol. 3, Paris, 1807, p. 143 and *Atlas*, plate 26, numbers 5 and 6 (drawings).

1058G. Spherical
Paris, France
Louvre Museum, no. MA 1775

Plate 11
Page 176

White marble - The corners and top left surface are damaged, but parts of eleven hour lines remain radiating from the gnomon hole to the lower edge of the spherical surface. Two day curves cross the hour lines. The large base is decorated with four steps, as well as with panels outlined with a narrow molding. A four-petalled flower can be seen in the center of both left and right sides.

Found at Lambèse (Lambaesis) in 1874 by the Mission Héron de Villefosse.

$H = 390$ mm
$W_{base} = 296$ mm
$D = 182$ mm

$tc_1 = 43$ mm
$c_1c_2 = 70$ mm
$c_2b = 50$ mm

(VI to VII)
$c_1 = 10$ mm
$c_2 = 28$ mm

$\beta = 41.3°$

Musée Africain du Louvre, p. 19, no. 254.

1059. Spherical
Cherchel, Algeria
Archeological Museum

H = 570 mm
W = 570 mm

A trace of lead fill is visible in the gnomon hole at the top center.

Marble - All eleven hour lines radiate from the gnomon hole to the lower edge of the dial face. Three day curves, which are not parallel to the lower edge, cross these hour lines. The dial base is decorated with two eight-leaf rosettes flanking a crescent moon. A half-cylinder molding separates these engraved figures from the lower sloping base.

Found near Cherchel in 1901.

Durry, Marcel, *Musée de Cherchel, Musées de L'Algérie*, Paris, 1924, p. 19.
Waille, Victor, "Rapport sur les Fouilles de Cherchel," *Revue Africaine* 46 (1902): 25-28.
Cf. *Revue Africaine* 47 (1903): plate 2 (photograph).

1060G. Curved
Varna, Bulgaria
Archeological Museum, no. 70

Plate 12
Page 176

A trace remains of a rectangular gnomon hole in the top surface of the dial.

White marble - The curved receiving surface of this dial is neither conical nor spherical. The oblique front section appears to be circular; the horizontal section, elliptical. The meridian line is curved with sagitta at right. Nine hour lines extend between two engraved solstices and cross an equinox. All trace of hour lines ten and eleven is broken away at right. Only the meridian extends from the gnomon hole to the summer solstice. The base of the dial is decorated with an uninscribed marble plaque. The dial and base lean backward 4.8° from normal. The sides and back are rounded to follow the curve of the shadow-receiving surface.

Found at Reka Devnia in the region of Trajan's Marcianopolis.

For $\varepsilon = 24°$

g = 128 mm
$\phi_W = 47°$
$\phi_S = 43°$

H = 415 mm
W = 2 x 153 mm
D = 196 mm

tw = 56.5 mm
we = 53 mm
es = 65 mm
sb = 17 mm

(VI to VII)
w = 15.5 mm
e = 25 mm
s = 35.5 mm

$\sigma = 49°$
$\gamma = 77°$

sagitta = 9 mm at equinox

Unpublished

11

12

1061G. Spherical
Vatican City
Vatican Museum, Galleria dei Candelabri

A gnomon hole is preserved in the top surface of the dial above the meridian.

Travertine marble - The rough-picked back of this dial is curved to follow the shape of the shadow-receiving surface. Eleven hour lines extend from the gnomon hole to the lower edge and are crossed by two parallel day curves.

Found at Ostia in 1858.

r_e = 129.9 mm

H = 278 mm
W = 320 mm
D = 215 mm
gw = 61 mm
we = 62 mm
eb = 127 mm
(VI to VII)
w = 17.5 mm
e = 34 mm
b = 47 mm
α = 75°

Unpublished

***

1062. Spherical
Volubilis, Morocco
Forum

White limestone - The dial is badly damaged but appears to have a spherical shadow-receiving surface. Hour lines extend from top to bottom of dial face.

H = 800 mm
W = 670 mm
D_{top} = 280 mm
tb = 400 mm

Measurements and descriptions from L. Janin.

1063G. Spherical Variant
Zagreb, Yugoslavia
Musée Archéologique, no. 205

Plate 13
Page 182

H = 684 mm
W = 637 mm
D = 251.5 mm
diameter of section = 505 mm

A 26 mm deep cylindrical gnomon hole (diameter 20 mm) is connected by a groove 57.5 mm long to the front of the dial.

Limestone - The shadow receiving surface of this dial is a spherical cap. Eleven hour lines radiate from the gnomon hole to the lower edge. No engraved day curves are visible. The Latin inscription includes letters IO and M in the the top corners and

SOLINVICTOCONSER
AVGNO

below the dial face.

Found near the church of Josipdol; possibly from Cakovac or Metulum near Ogulin. Presented to the museum in 1882.

Hour lines six to ten have the following lengths from gnomon to base: 528 mm, 517 mm, 487 mm, 420 mm, 314 mm. The arcs separating the hour lines along the base of the shadow-receiving surface from the second hour line to the eleventh hour line have the following lengths: 124, 124, 109, 115, 103.5, 111.5, 112.5, 155, 154 mm.
The angles between hour lines from the first to the eleventh hour line are: 12°, 21°, 11.5°, 12.5°, 14°, 12.5°, 13°, 13.5°, 17°, 20°.

Brunsmid, Josip, *Kameni Spomenici*, Zagreb, 1904-11 pp. 113-14.
CIL, III, 3020.
Cf. Frankfurter, S., AEM 8(1884): 166.

1064. Unfinished Spherical
Present location unknown

White marble - A vertical back, horizontal top and sloping front characterize the general shape of this dial. Two schematic feet flank its base. A rough-picked area divided by a vertical strip, and bounded by a semicircle, marks the location of the unfinished shadow-receiving surface.

H = 330 mm
W = 470 mm

Found in the Agora of the Italians, Delos, Greece.

Deonna, p. 191 and pl. 543 (photograph).

***

1065. Unfinished Spherical
Bucharest, Rumania
National Archeological Museum

White marble - The sundial is undamaged, except for chipped upper corners; and yet, no gnomon hole is preserved. Also there are no visible hour lines nor any visible day curves. The back is cylindrical.

H = 290 mm
W = 210 mm
r = 105 mm
$\alpha = 75°$

Found in Constanta, Rumania.

Ionescu-Carligel [1], p. 203.

1066G. Hemispherical Variant
Paris, France
Louvre Museum, no. AO 3025

A hole at the junction of the summer solstice and the meridian, 11 mm x 10 mm and 18 mm deep, connects via a trough with the lower edge of the spherical surface. Also, there is evidence of a groove and a depression in the top surface of the monument above the meridian on the dial surface.

radius = 75 mm

tw = 51.5 mm
we = 18.5 mm
es = 41 mm

(VI to VII)
w = 14.5 mm
e = 17 mm
s = 20 mm

Dark stone - A spherical depression in this large hexagonal monument has been marked for use as a sundial. Nine hour lines are preserved (3,4,5,6,7,8, 9,10,11) and three engraved day curves are visible. None of the hour lines continue beyond either solstice. A Nabataean inscription on the monument dates from the year 94 A.D.

Found at Cippe de Dmeir, near Damascus, Syria.

Clermont-Ganneau, *Recueil Orientale I*, Paris, 1888, p. 48.

1067G. Hemispherical — Plate 14
Stuttgart, Germany — Page 182
Württembergisches Landesmuseum, no. R193,3

There is no trace of a gnomon hole in the top edge. Reconstruction in the area of the bottom of the dial has obliterated any traces of a gnomon there.

Pumice stone - The hemispherical surface has been cut out of a cylindrical block. A square protrusion (for mounting?) 80 mm x 80 mm remains in the center of the bottom surface. A Latin inscription on the lateral surface of the cylinder reads:

-]NIVS•TA[-

The winter solstice seems to be punctuated with dots at the points where it is joined by the hour lines.

Found in Cannstatt in 1843.

H = 75 mm
$\text{diameter}_{\text{top}}$ = 154 mm
$\text{diameter}_{\text{base}}$ = 184 mm
tw = 26.5 mm
ws = 92+ mm
(VI to VII) w = 14 mm
(II to III) s = 23 mm

Dated from the artwork of a bronze figurine found nearby to the second century A.D.

Reconstruction makes determination of intended latitude difficult. Schlieben, who examined the unrestored pieces, finds $\phi = 42°30'$ for r = 80 mm. He completes the inscription as follows: LICINIVS TARENTINVS.

Brambach, G., *Corpus Inscriptionum Rhenanarum*, 1867, p. 294, no. 1578 (CIL, XIII, 6444).
Haug, F., and Sixt, G., *Die Römischen Inschriften und Bildwerke Württembergs*, Stuttgart, 1914, pp. 424-25 no. 298 (photographs).
Paulus, *Jahrbücher des Vereins von Altertumsfreunden im Rheinlande (Bonner Jahrbücher)* 4(1844): 90-91 and plates i and ii (drawings).
Schlieben, "Römische Sonnenuhren in Wiesbaden und Cannstadt," AVN 20(1888): 316-33 and table 13, figures 7a and b (drawings).

13

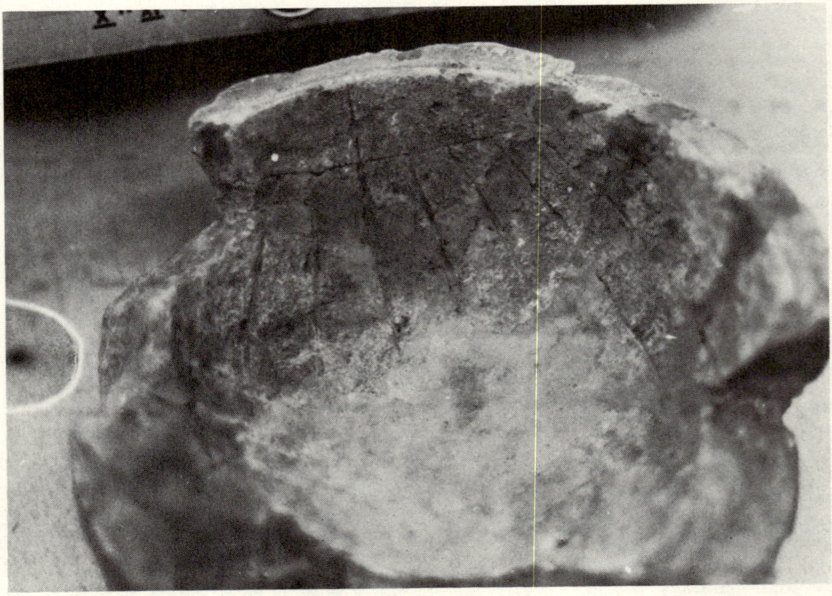

14

1068G. Hemispherical — Plate 15, Page 188
Vatican City
Vatican Museum, Galleria dei Candelabri, no. II 90
2439

There is no trace of a gnomon hole in the top edge of this dial. The rough surface at the front suggests that the face was once a complete hemisphere. If this was the case, then the gnomon could have been mounted in the bottom of the dial as S. Piale indicates, or in the top edge of the portion which is now broken away.

Luna marble, white - In addition to eleven hour lines and seven day curves (marking the entrance of the sun into each of the signs of the zodiac), a circle with vertical diameter coinciding with the meridian and equal to the distance between the solstices, and an isosceles triangle with axis also equal to the distance between the solstices and coinciding with the meridian, have been engraved on the shadow-receiving surface. An inscription listing the month names and the first letters of the names of the signs of the zodiac, all in fine Greek characters, is distributed among the day curves in the order below:

H = 256 mm
W = 513 mm
D = 230 mm

tw = 95 mm
wc_1 = 14 mm
c_1c_2 = 34 mm
c_2e = 45 mm
ec_3 = 45 mm
c_3c_4 = 34 mm
c_4s = 14 mm
sb = (57 mm)

(VI to VII)
w = 40.5 mm
c_1 = 43.5 mm
c_2 = 50 mm
e = 59 mm
c_3 = 64 mm
c_4 = 67 mm
s = 96 mm

ιανουαριος	τοξ(ότης)	αἰγ(όκερως)	δεκεμβριος
φεβραριος	σκό(ρπιος)	ὑδρ(οχόος)	νοεμβριο[s]
[μ]αρτιος	ζυγ(ós)	ἰχθ(ύες)	οκτωβρι[os]
[απ]ριλιος	παρθ(ένος)	κρει(ός)	σεπτεμβ[ριος]
[μα]ιος	λέων	ταῦρ(ος)	αυγο[υστος]
[ιουν]ιος	καρκί(νος)	δίδυμ(οι)	ι[ουλιος]

Found in the Via Palombara on the Esquiline in 1805.

The triangular figure represents the increase in daylight from winter solstice to summer solstice. The circular figure is the *menaeus* described by Vitruvius. Other dials with daylight triangles include 4001G, 1044, 3046.

1068G. continued

r_e = 225 mm
$\varepsilon = 23°36'$
$\phi = 42°$
arc c_1 represents the path of the sun with declination 20°6'
arc c_2 represents the path of the sun with declination 11°30'
arc c_3 represents the path of the sun with declination -20°6'
arc c_4 represents the path of the sun with declination -11°30'

IG, XIV, 1307 (drawings) and CIG, III, 6179 (drawings).
Lippold, G., *Die Skulpturen des Vaticanischen Museums, Galleria dei Candelabri*, vol. 2, Berlin, 1956, p. 223, no. 90 and table 102 (photograph).
Piale, S., *Memorie Enciclopediche Romano*, vol. 5, 1805, pp. 102-09 (drawings).
cf. Diels, pp. 168-70 (drawings).
Gatty, pp. 34-35 (drawing).

***

1069G. Hemispherical
Bergama, Turkey
Archeological Museum, portico

White marble - Grids have been engraved on both the north and south facing halves of the hemispherical shadow-receiving surface of this dial. Since gnomon holes near the top edge may be associated with each dial grid, the hole in the bottom center of the hemisphere presumably served to drain off accumulated water.

H = 223.5 mm
diameter of hemisphere = 323 mm

diameter of drain hole = 13 mm

Each dial face is treated separately in the description which follows:

1069G. continued

South Face

Eleven hour lines extend from the winter solstice across the equinox to the summer solstice. The meridian continues to the top edge.

tw = 20 mm
we = 49 mm
es = 48.5 mm
(VI to VII)
w = 18.8 mm
e = 32.2 mm
s = 40 mm

For $\varepsilon = 24°$, a graphic determination of the gnomon length gives g = 92.4 mm.

Arc lengths along summer solstice from the sixth to the eleventh hour lines measure: 40, 43.5, 43.5, 40, 49 mm.

North Face

A spike gnomon 47 mm long is fixed along the meridian 23.5 mm from the top edge.

tg = 23.5 mm
gs = 103 mm
se = 109 mm
ew = 123.5 mm

Eleven hour lines extend from the summer solstice across the equinox to the winter solstice. The meridian continues to the top edge.

(VI to VII)
w = 38 mm
e = 54 mm
s = 62.2 mm

For $\varepsilon = 24°$, a graphic determination of the gnomon length gives g = 21 mm.

Arc lengths along the summer solstice from the sixth to the eleventh hour lines measure: 38, 36, 37, 24.8, 22.1 mm. Arc lengths along the winter solstice from the sixth to the twelfth hour lines measure: 62.2, 40.3, 47, 42, 47, 39 mm.

On both faces, for $\varepsilon = 24°$, $\phi = 45°$.

Rehm, A., "Zwillingssonnenuhr aus Pergamon," Ath. Mitt. 36(1911):251-268 with photograph (Diels, pp. 165-167).
Cf. Suheyl-Unver, "Sur les cadrans solaires horizontaux et verticaux en Turquie," *Archives Internationales d'Histoire de Science* 7(1954): figs. 14 and 15 (photos).

1070G. Hemispherical
Delos, Greece
Museum, portico storage room

White marble - Only the left half of this dial is preserved. However, broken edges suggest that the shadow-receiving surface was once a complete hemisphere. Six finely engraved hour lines and three day curves are visible. The hour lines do not extend beyond the solstice curves.

H = 287 mm
W = 156 mm
D = 225 mm
tw = 83 mm
we = 63 mm
es = 64 mm
(III to IV)
w = 31 mm
e = 41 mm
s = 45 mm

r_m = 150 mm
r_e = 156 mm
ϕ_w = 31°
ϕ_s = 35°

Deonna, p. 191.

1071G. Hemispherical
Delos, Greece
Museum, portico, no. B2929-8545

Plate 16
Page 188

There is no trace of a gnomon hole in the smooth surface at the top or the broken surface at the lower front.

White marble - The broken edge at the front of this dial suggests that the dial face was once a complete hemisphere. Now traces of eleven hour lines remain on worn shadow-receiving surface. Where the surface is not as badly worn hour lines can be seen to extend beyond the lowest day curve to the broken edge. In all, four day curves are preserved. An engraved line in the top surface marks the meridian position. The diameter of the spherical hollow is approximately 200 mm.

r_e = 113 mm
$\overline{\phi}$ = 53°
ϕ = 37°

c_1 corresponds to δ = 24°48'
c_2 corresponds to δ = 22°12'
c_3 corresponds to δ = 14°30'

Dials 1070G and 1068G are also probably broken hemispheres.

H = 79 mm
W = 255 mm
D_{max} = 121 mm

tc_1 = 55.5 mm
c_1c_2 = 5 mm
c_2c_3 = 15 mm
c_3c_4 = 28 mm
c_4l = 54 mm

(IX to X)
c_1 = 19.5 mm
c_2 = 20.5 mm
c_3 = 24 mm
c_4 = 29 mm

Deonna, p. 190, pl. 547 (photograph).

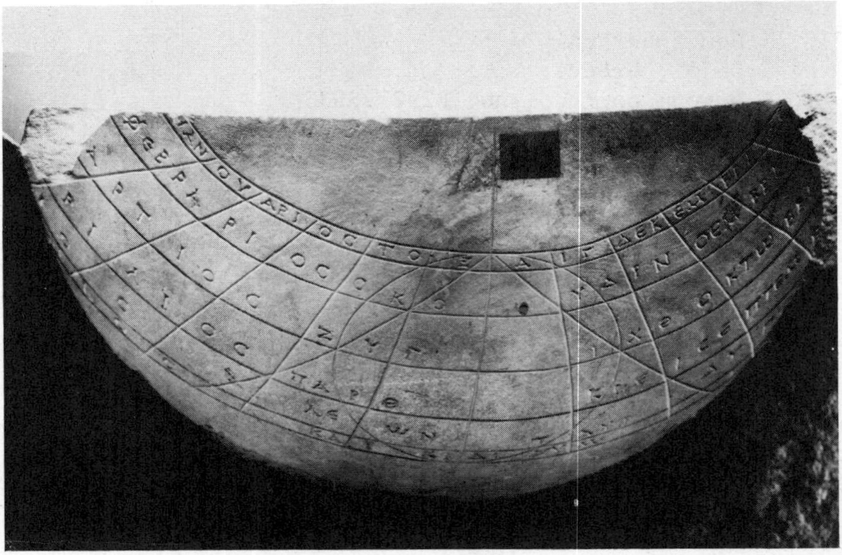

15

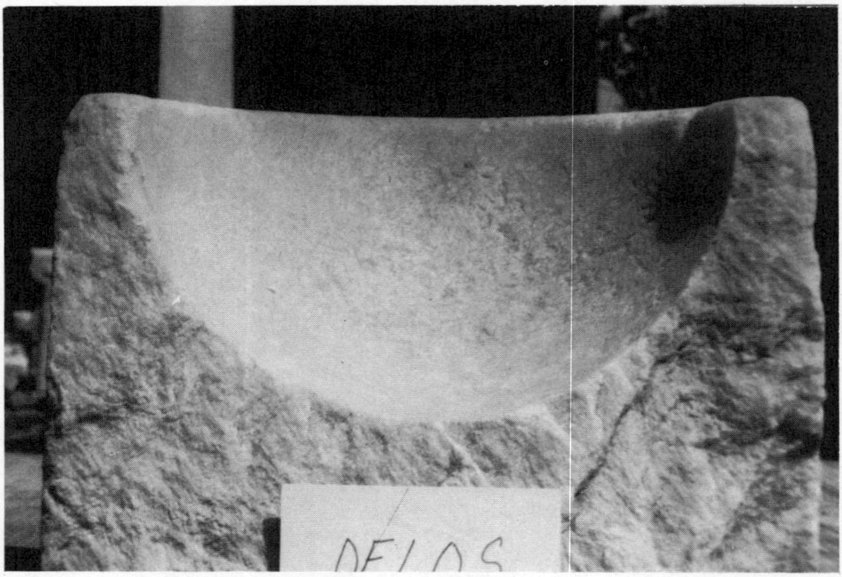

16

1072G. Hemispherical
Delos, Greece
Museum, storage room, no. B4367

Plates 17, 18
Page 190

A probable gnomon hole can be seen in the top surface of the dial. It is cylindrical, 22 mm in diameter, and 44 mm deep.

H = 448 mm
W = 317 mm
D = 192 mm

ts = 282.5 mm
se = 66 mm
ew = 63 mm
wb = 69 mm

(VI to VII)
w = 28 mm
e = 39 mm
s = 42 mm

White marble - The transposed halves of a hemisphere form the shadow-receiving surface of this dial. On each side five hour lines stretch between the two solstices and across the equinox. The meridian line is not engraved on either side. An engraved line connects the equinoxes across the top of the dial. On the west facing side, the summer solstice continues from the hollow onto the plane surface. On the east facing side, an engraved line cuts the plane meridian edge a little south of the summer solstice. Greek inscriptions mark the day curves.

East facing side:
τροπαὶ
ἰσημερία
τροπαὶ θεριναί

West facing side:
τροπαὶ
ἰσημερία
[τροπαὶ] θεριν[αί]

The prow of a ship, pierced by a small hole, decorates the south side of the block.

For $\varepsilon = 24°$, $r_m = 150$ mm, $r_e = 149$ mm, $\phi_W = 36°$, $\phi_S = 32°$.
Other transposed spherical dials include 1073 and 1074. A conical dial from Sparta, 3107G, also has a base shaped like the prow of a ship.

BCH 36(1912): 393, no. 8 and plate 5 (photograph).
Deonna, p. 192, figure 229 (drawing) and plates 548 and 549 (photographs).

17

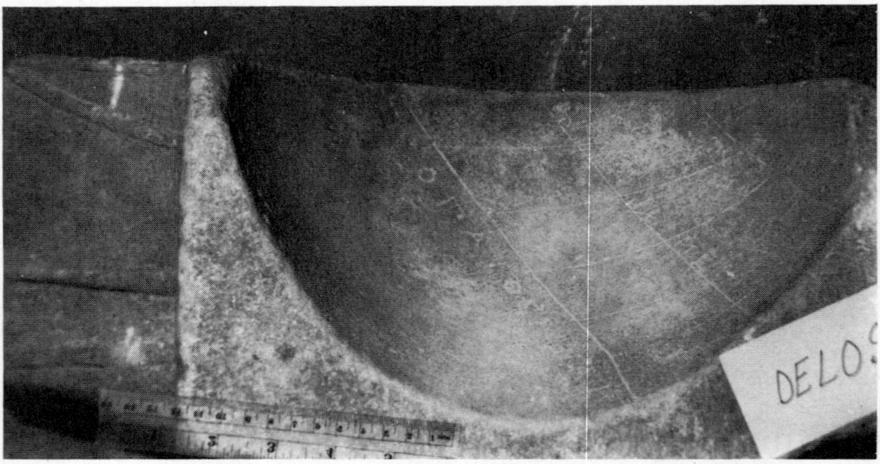

18

1073. Hemispherical
Present location unknown

H = 330 mm
W = 260 mm

White marble - The sketch published in *Inscriptiones Graecae* was drawn from photographs. The equinox, the summer solstice line and a line near the summer solstice are marked with Greek inscription.

East facing side:
ἰσημερία
κύων ἐκφανής
τροπὴ θερινή

West facing side:
ἰσημερία
κύων ἐκφανής
τροπὴ θερινή

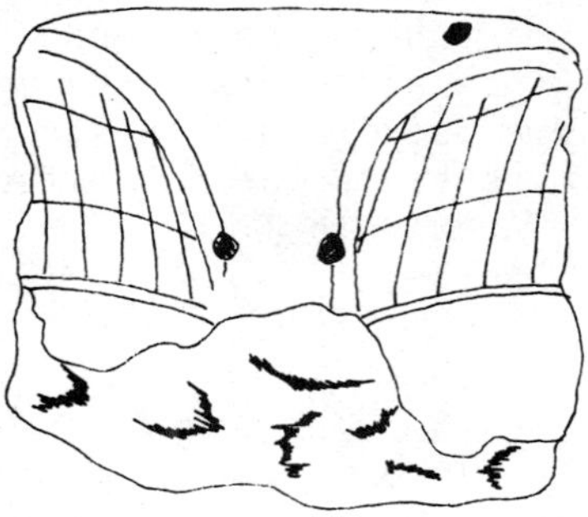

The dial is composed of transposed quarter spheres each engraved with five hour lines and four day curves.

Found along a street on the island of Kea.

A similar transposed hemispherical dial was found recently on the island of Samothrace by James McCredie. (See catalogue number 1075.)

IG, XII, 5.1, 645 and p. 334 (drawing).

1074. Hemispherical Variant
Present location unknown

H = 450 mm
W = 300 mm

Limestone - Ingholt's photo shows half of a hemisphere with part of a supplementary dial south of the main cavity. The main dial face is broken along the meridian, giving the impression that a complete hemisphere was intended. Six hour lines and parts of seven day curves are preserved. The top western face bears the following inscription:

Ἀπελλᾶι(ος)
Δειõς
Ὑπερβερε(ταῖος)
[Γ]ορπιᾶιος
Λῷος
[Π]άνημος

The summer solstice is engraved:

θερινός

The equinox is engraved:

ἰσ[η]με[ρινός]

The winter solstice is engraved:

[χειμ]εριν[ός]

At least four hour lines and a day curve are visible on the supplementary dial. No inscription is preserved.

Found in a house inside the temple of Bêl at Palmyra, March, 1925.

Ingholt, H., "Palmyra," *Berytus* 3(1936):112-14 (photograph).

1075. Hemispherical
Samothrace, Greece
Excavation Museum, no. 70.894

Photographs of this dial show a dark brown stone inscribed with lines, curves, and Greek letters. The shadow-receiving surface appears to be transposed hemispheres similar to 1072G. As many as seven day curves cross the four preserved hour lines on both faces. One of these day curves is inscribed

ἰσημερινός.

Remaining inscription is difficult to decipher. A segment of the equinoctial line between hour points is 39 mm long.

Information from James McCredie, Samothrace Excavations.

Roofed Spherical Dials

Measured Quantities

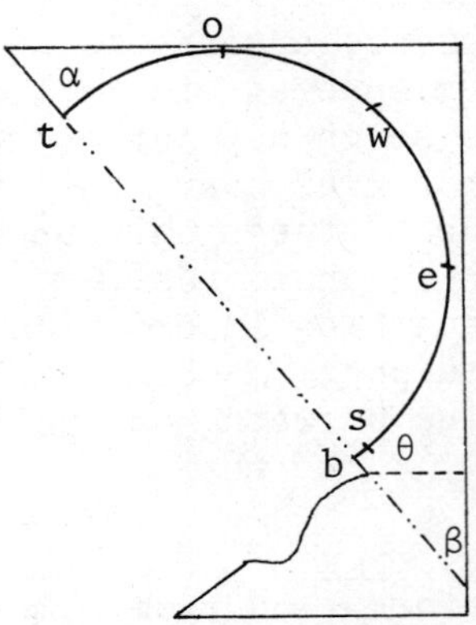

H = height
W = width
D = depth
d = diameter of front opening

tw = distance along meridian from top front edge of dial to winter solstice curve
we = distance along meridian from winter solstice curve to equinox curve
es = distance along meridian from equinox curve to summer solstice curve
sb = distance along meridian from summer solstice curve to bottom edge of dial face

(VI to VII)
w = distance along winter solstice curve from the sixth to the seventh hour point
e = distance along equinox curve from the sixth to the seventh hour point
s = distance along summer solstice curve from the sixth to the seventh hour point

α = acute angle between oblique front of dial and horizontal
β = acute angle between oblique front of dial and vertical
θ = obtuse angle between oblique front of dial and horizontal

Computed Quantities

ϕ = intended latitude
ε = obliquity of ecliptic
r = radius of spherical dial surface
oe = diameter of equinoctial day curve

2001G. Spherical Variant
Altino, Italy
Museo di Altino, no. AL10

Plate 19
Page 198

Marble of Verona - The spherical shadow-receiving shell is tipped up on end between two antae at an angle of about 27° from vertical. Although the top and sides are broken away, hour lines 3 through 9 can be seen extending from the winter solstice across the equinox to the summer solstice. A boat has been carved in high relief on the base of the dial.

H = 680 mm
W = 415 mm
D = 275 mm

we = 134 mm
es = 129 mm

(VI to VII)
w = 22 mm
e = 58 mm
s = 98 mm

$\beta = 27°$

Found in 1931 during agricultural work in the Altino area.

For $\varepsilon = 24°$, $r_W = 160$ mm
$\phi_W = 46°$, for $(\text{VI to VII})_e = 58$ mm
$= 43°$, for $(\text{VII to VIII})_e = 61$ mm
$r_S = 154$ mm
$\phi_S = 44°$, for $(\text{VI to VII})_e = 58$ mm
$= 41°$, for $(\text{VII to VIII})_e = 61$ mm

The bowl shape of this dial is similar to that of 2019G from Pompeii. Two other dials in the catalogue have ship-shaped bases: 3107G from Sparta, and 1072G from Delos.

Unpublished

2002G. Spherical Variant
Aquileia, Italy
Via Sacra, on brick column
near Porto Fluviale Romano

White marble - The slightly rounded top surface of this dial is pierced by an irregularly shaped hole. Four small human profiles decorate the upper and lower corners of the oblique front surface. A wide lip surrounds the circular front opening. All eleven engraved hour lines and day curves remain on the dial surface which is supported by a cyma reversa base.

H = 717 mm
W = 585 mm
D = 460 mm
th_1 = 148 mm
h_1h_2 = 56 mm
h_2w = 156.5 mm
we = 223 mm
es = 218 mm
sb = 24 mm
(V to VI)
w = 41.5 mm
e = 95 mm
s = 158 mm
$\theta = 111°$
$\alpha = 58°$

For $\varepsilon = 24°$, r_w = 266 mm
$\phi_w = 47°$, for $(V \text{ to } VI)_e$ = 95 mm
$= 42°$, for $(IV \text{ to } V)_e$ = 104 mm
r_s = 260 mm
$\phi_s = 46°$, for $(V \text{ to } VI)_e$ = 95 mm
$= 40°$, for $(IV \text{ to } V)_e$ = 104 mm

Dial 2005G, also found at Aquileia, is quite similar to this dial without relief decoration.
See 2017G for definitions of h_1 and h_2.

Bracchi, col. 65-66, fig. 9 (photograph).
cf. Brusin, Giovanni, *Gli Scavi di Aquileia*, Udine, 1934, p. 115, no. 10, item 17 (photograph).

2003G. Spherical Variant
Aquileia, Italy
Archeological Museum, garden, no. 1097

Plate 20
Page 198

White marble - The slightly rounded top is damaged at the center. The hour lines are symmetric about the meridian. Two four-leafed rosettes decorate the top corners; two symmetric volutes and stylized legs decorate the base. A 23 mm wide relief lip borders the front circular section of the dial face. The meridian line continues beyond the summer solstice onto the lip.

Found in Monastero near the east gate of Admiral Ritter's Villa Vincentina.

Hour arcs along the winter solstice from the third to the sixth hour lines measure: 30, 18, 30 mm.

H = 453 mm
W = 350 mm
D_{top} = 260 mm
D_{base} = 184 mm
d = 304 mm

tw = 63 mm
we = 132 mm
es = 134 mm
sb = 15 mm

(III to IV)
w = 30 mm
e = 61 mm
s = 82 mm
(IV to V)
w = 18 mm
e = 61 mm
s = 88 mm
(V to VI)
w = 30 mm
e = 63 mm
s = 99 mm

$\alpha = 66°$
$\beta = 24.9°$

For $\varepsilon = 24°$, r_W = 158 mm

$\phi_W = 40°14'$, for $(\text{V to VI})_e$ = 63 mm

$\phi_W = 42°$, for $(\text{IV to V})_e$ = 61 mm

This dial might have been made for use in Rome. See catalogue entry 2022, for a similar roofed spherical dial found on the Palatine Hill in Rome.

Bracchi, p. 55, fig. 5 (photograph) and p. 67, note 5.
Kenner, p. 4, figs. 3 and 4 (drawings), pp. 5-6.
Cf. Gatty, p. 33.

19

20

2004G. Spherical Variant
Aquileia, Italy
Archeological Museum, portico, no. 3497

The groove in the rough upper surface of the dial is certainly not part of the original design.

White marble - Its heavily decorated base is the distinctive characteristic of this dial. Large amphorae have been engraved into the center of leaf-framed squares on three sides. Above, an acanthus leaf base supports the shadow receiving hollow. Two three-leaf rosettes are visible at the left and right front just below the lip. Seven hour lines (3,4,5,6,7,8,9) and the equinox and summer solstice are preserved. The "ceiling" of the dial is broken away at the sides and top.

base:
H = 695 mm
W = 440 mm
D = 430 mm

dial:
H = 288 mm
W = 358 mm
D = 330 mm

te = 130 mm
es = 141 mm
sb = 12.5 mm

(VI to VII)
e = 57 mm
s = 81 mm

$\beta = 29°$

For $\varepsilon = 24°$, $r_s = 168$ mm
$\phi_s = 42°$, for $(VII \text{ to } VIII)_e = 65$ mm
$\phi_s = 50°$, for $(VI \text{ to } VII)_e = 57$ mm

Bracchi, col. 60-62.
Kenner, pp. 2-3, fig. 2 (drawing).

2005G. Spherical Variant
Aquileia, Italy
Archeological Museum, no. 1052

Gray stone - Although the top and right side of this dial are broken away, nine hour lines are preserved (2,3,4,5,6,7,8,9,10) as well as part of three day curves. The grooves are 9.5 mm wide. The dial is supported by a simple cyma reversa base. A wide lip surrounds the circular section formed by the oblique front plane.

H = 735 mm
W = 599 mm
D = 404 mm

tw = 140 mm
we = 232 mm
es = 237 mm
sb = 45 mm

(VI to VII)
w = 36 mm
e = 97 mm
s = 163 mm

θ = 110°
α = 66°

Hour arcs along the summer solstice from the third to the sixth hour lines measure: 153, 158, 163 mm.

For $\varepsilon = 24°$, $r_W = 277$ mm, $r_S = 283$ mm
$\phi_W = 48°$, $\phi_S = 49°$

Note that this dial is quite similar to the dial with relief decorations set up along the Via Sacra in Aquileia (2002G).

Bracchi, col. 67, note 5.

2006G. Spherical Variant
Aquileia, Italy
Archeological Museum, no. 1088

Plates 21, 22
Page 202

White marble - The grid of hour lines and day curves forming this dial has been engraved on the inner surface of a large marble flower sprouting from a marble stem and calyx. A series of petals surrounds the circular opening on the oblique front. Parts of two curving tendrils, sprouting from a central point below the meridian, can be seen carved into the front surface of the dial. Eight hour lines (3,4,5,6,7,8,9,10) and parts of three day curves are preserved.

H = 460 mm
W = 390 mm
D = 300 mm

tw = 66 mm
we = 147 mm
es = 150 mm
sb = 18.5 mm

(V to VI)
w = 32 mm
e = 64 mm
s = 10.8 mm

$\theta = 119°$

For $\varepsilon = 24°$, $r_w = 175$ mm

$\phi_w = 40°30'$, for $(IV\ to\ V)_e = 70$ mm

$\phi_w = 45°50'$, for $(V\ to\ VI)_e = 64$ mm

$r_s = 179$ mm

$\phi_s = 41°43'$, for $(IV\ to\ V)_e = 70$ mm

$\phi_s = 47°$, for $(V\ to\ VI)_e = 64$ mm

This dial may possibly be a copy of 2011G now in the Archeological Museum in Udine.

Bracchi, col. 52, figs. 3 and 4 (photographs).

21

22

2007G. Spherical Variant
Aquileia, Italy
Archeological Museum, no. 844

White marble - The rounded top of this dial is partially broken away, but nine hour lines (2,3,4,5,6,7,8, 9,10) and most of all three day curves are preserved. Two four-leaf rosettes decorate the top left and right corners; two tendrils decorate the bottom left and right corners. The dial is supported by a lesbian cyma base. A wide lip surrounds the circular segment cut off by the oblique front plane.

Hour arcs along the summer solstice from the fourth to the seventh hour lines measure: 135, 145, 140 mm.

For $\varepsilon = 24°$, $r_w = 286$ mm and $r_s = 280$ mm
$\phi_w = 48°11'$ and $\phi_s = 47°$.

H = 635 mm
W_{base} = 543 mm
W_{top} = 670 mm
D = 321 mm
d = 510 mm
tw = 64 mm
we = 240 mm
es = 235 mm
sb = 19 mm
(V to VI)
w = 35 mm
e = 100 mm
s = 145 mm
$\alpha = 71°$
$\theta = 101°$

Bracchi, col. 67, note 5.

***

2008G. Spherical Variant, fragment
Aquileia, Italy
Museum, study collection, no. 47848

Porous stone - Parts of three hour lines and two day curves are preserved. All incisions are deep and pointed at the bottom. A relief lip surrounds the front opening.

(II to III)
e = 46 mm
s = 76 mm
es at third hour
= 103 mm

Unpublished

2009G. Spherical Variant, fragment
Aquileia, Italy
Museum, study collection

Fine grained stone - Parts of two hour lines, equinox, and summer solstice are preserved. The light incisions are rather crude. A relief lip surrounds the front opening.

(II to III)
e = 39 mm
s = 51 mm

The arcs between equinox and summer solstice along the second and third hour lines measure: 40 mm and 58 mm.

Unpublished

***

2010G. Spherical Variant, fragment
Aquileia, Italy
Museum, study collection, no. 53625

White stone - Parts of four hour lines and the winter solstice are preserved. The deep incisions are pointed at the bottom but differ from those on 2008G.

(III to IV)
w = 25 mm
(IV to V)
w = 18 mm
(V to VI)
w = 25 mm

Unpublished

2011G. Spherical Variant, fragment
Udine, Italy
Civic Museum, no. 174

Calcareous stone from Carso - Five fragments have been joined together in an unfortunate restoration to form the dial surface supported by an elaborate base. In all, parts of five hour lines, including the meridian, and the summer solstice are visible. Petals decorate the outer edge of all fragments. On the tall base a woman standing in front of curling tendrils, which spring from a point below the meridian, has been carved in high relief. The woman's left hand rests on a herm. Curling acanthus leaves frame the central figures.

H = 930 mm
W = 340 mm
D = 280 mm

(VI to VII)
s = 159 mm

Found in the necropolis zone of Aquileia.

Bracchi notes the similarity between this dial and 2006G.

Bracchi, cols. 49-51, figs. 1 and 2 (photographs).
Kenner, pp. 6-7 (drawings).

2012G. Spherical Variant
Pula, Yugoslavia
Archeological Museum (Arheoloski Muzej Istre)

Limestone - The top and sides have been restored to this spherical variant which originally consisted of a fragment preserving seven hour lines (3,4,5,6,7,8,9) and three engraved day curves. The restored circular section in the oblique front of the dial has diameter 387 mm. A small portion of an 11 mm lip is still visible.

$H = 664$ mm
$W = 485$ mm
$D_{top} = 416$ mm
$D_{base} = 360$ mm
$tw = 247$ mm
$we = 159$ mm
$es = 159$ mm
$sb = 46$ mm

(VI to VII)
$w = 28$ mm
$e = 90$ mm
$s = 119$ mm
$\theta = 117.5°$
$\alpha = 63.5°$

For $\varepsilon = 24°$, $r = 190$ mm
$\phi = 64°55'$, for $(\text{VI to VII})_e = 90$ mm
$\phi = 52°39'$, for $(\text{VII to VIII})_e = 79$ mm
$\phi = 45°36'$, for $(\text{VIII to IX})_e = 71$ mm

Unpublished

2013G. Spherical Variant
Pula, Yugoslavia
Archeological Museum

A keyhole-shaped light hole in a rectangular depression (21 mm wide x 36 mm long) is now visible in the restored top surface of the dial. The center of the restored hole is 225 mm from the left edge and 102 mm from the front.

Limestone - Museum officials have completely restored this dial starting with a fragment on which seven hour lines (3,4,5,6,7,8,9) and three day curves were preserved. A 10 mm margin follows the circular section (d = 342 mm) and the square outer boundary of the oblique front.

Found near Pula in 1940.

$H = 503$ mm
$W = 441$ mm
$D_{top} = 345$ mm
$D_{base} = 242$ mm
$to = 88$ mm
$ow = 132.5$ mm
$we = 141$ mm
$es = 145$ mm
$sb = 13$ mm
(VI to VII)
$w = 25$ mm
$e = 63$ mm
$s = 97.5$ mm
$\theta = 125°$
$\alpha = 55°$

For $\varepsilon = 24°$, $r_w = 168$ mm, $r_s = 173$ mm
$\phi_w = 44°22'$, for $(\text{VIII to IX})_e = 61$ mm
$\phi_w = 49°29'$, for $(\text{VII to VIII})_e = 67$ mm
$\phi_w = 43°48'$, for $(\text{VI to VII})_e = 63$ mm
$\phi_s = 45°58'$, for $(\text{VIII to IX})_e = 61$ mm
$\phi_s = 47°40'$, for $(\text{VII to VIII})_e = 67$ mm
$\phi_s = 42°18'$, for $(\text{VI to VII})_e = 63$ mm

Cf. *Archäologischer Anzeiger* 57 (1942): col. 286.

2014G. Spherical Variant
Trieste, Italy
Civic Museum of History and Art

Limestone - The top, sides, and front of this dial are damaged, but seven hour lines (3,4,5,6,7,8,9) and parts of three day curves are preserved. On the front surface of the dial two narrow, curving leaves meet below the meridian. The base has a cyma profile.

H = 470 mm
W = 306 mm
D_{base} = 160 mm
we = 112 mm
es = 112 mm
(VI to VII)
w = 15 mm
e = 51 mm

θ = 129°

For ε = 24°, r = 134 mm
ϕ = 48°55', for $(V \text{ to } VI)_e$ = 46 mm
ϕ = 43°13', for $(VI \text{ to } VII)_e$ = 51 mm
ϕ = 46°9', for $(VII \text{ to } VIII)_e$ = 48.5 mm

Unpublished

2015. Spherical Variant
Trieste, Italy
Civic Museum of History and Art, magazine

Limestone - Five hour lines (6,7, 8,9,10) and three day curves are preserved. Parts of an acanthus leaf decoration decorate the upper and lower corners of this preserved right front face. A relief margin emphasizes the front circular opening.

Found in the ruins of a Roman fulling establishment at San Sabba near Trieste.

$H = 280$ mm
$H_{oblique} = 320$ mm
$r = 152$ mm
we = 122 mm
es = 128 mm
sb = 14 mm
$(VI \text{ to } VII)_e = 56$ mm

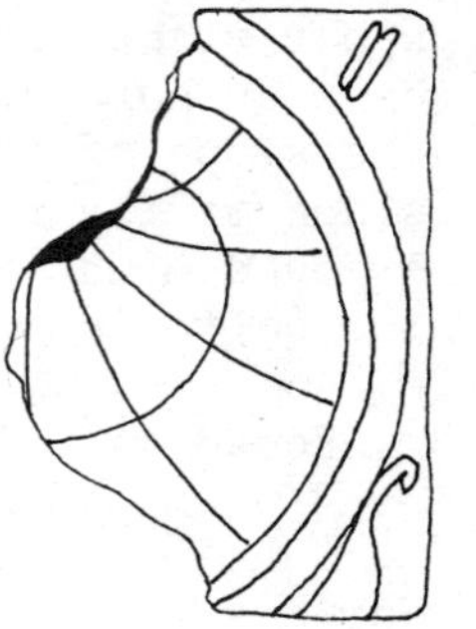

For $\varepsilon = 24°$, $r_w = 146$ mm
$\phi_w = 42°45'$, for $(VI \text{ to } VII)_e = 56$ mm
$= 36°52'$, for $(VII \text{ to } VIII)_e = 61$ mm
$= 43°50'$, for $(VIII \text{ to } IX)_e = 55$ mm
$r_s = 153$ mm
$\phi_s = 45°35'$, for $(VI \text{ to } VII)_e = 56$ mm
$= 40°19'$, for $(VII \text{ to } VIII)_e = 61$ mm
$= 46°34'$, for $(VIII \text{ to } IX)_e = 55$ mm

Grablovitz, Giulio, "Frammento d'Orologio Solare Rinvenuto a S. Sabba," *Archeografo Triestino* 12 (1886): 401-10, figs. 1 and 2 (drawings).
Puschi, A. "Relazione Intorno alle Scoperte Archeologiche di San Sabba," *Archeografo Triestino* 12 (1886): 389-400.

2016G. Spherical Variant
Split, Yugoslavia
Arheoloski Muzej

Travertine marble - Remains of a broken top suggest that it was slightly rounded. Nine hour lines (3,4,5,6,7,8,9,10,11) are preserved, extending from the winter solstice across the equinox to the summer solstice. A 24 mm lip surrounds the circular section in the oblique front. A tiny horn decorates the upper right corner. A stepped cyma base supports the spherical segment. The slope of the front follows the slope of the summer solstice curve. A round mount hole is preserved in the bottom surface.

Found at Salona.

H = 480 mm
W = 454 mm
D_{top} = 353 mm
D_{base} = 224 mm
d = 406 mm

we = 200 mm
es = 207.5 mm
sb = 8 mm

(VI to VII)
w = 45 mm
e = 89 mm
s = 107 mm

$\theta = 115°$
$\alpha = 65°$

For $\varepsilon = 24°$, r_w = 239 mm

ϕ_w = 44°40', for $(\text{VI to VII})_e$ = 89 mm
= 43°27', for $(\text{VII to VIII})_e$ = 86 mm
= 49° 7', for $(\text{VIII to IX})_e$ = 94.5 mm

r_s = 248 mm

ϕ_s = 43°20', for $(\text{VI to VII})_e$ = 89 mm
= 41°32', for $(\text{VII to VIII})_e$ = 86 mm
= 43°46', for $(\text{VIII to IX})_e$ = 94.5 mm

Unpublished

2017G. Spherical Variant
Volterra, Italy
Museo Etrusco Guarnacci

Plates 23, 24
Page 212

A green-stained rectangular (66 mm x 72 mm) depression surrounds a vaguely elliptical hole (44 mm x 35 mm) in the top surface of the dial.

White marble - The diameter of the circular section cut off by the oblique front plane of the dial is 260 mm. The equinox and solstices converge toward the hole at top. Eleven hour lines extend between the winter and summer solstices in a pattern symmetric about the meridian. Two six-leaf rosettes inscribed in circles decorate the upper left and right corners. The lower left and right corners are supported by feet ending in four nobby, widely-spaced toes. A three line Latin inscription remains inscribed in the back of the dial.

Q·POEN[A]
APER·IIII·VIR
AED·D·S·P·F·C
(Aedilis de sua pecunia faciendum curavit)

For r_s = 131 mm, oe = 191 mm,
$\phi = 43°$

The lion feet on this dial are quite similar to those on 2021G in the Vatican Museum.

H = 339 mm
W = 317 mm
D_{top} = 265 mm
D_{bottom} = 170 mm
th_1 = 69 mm
h_1h_2 = 35 mm
h_2w = 96 mm
we = 108 mm
es = 110 mm
sb = 4 mm
(VIII to IX)
w = 19 mm
e = 52 mm
s = 74 mm
$\alpha = 56°$
$\theta = 125°$

(h_1 = front of top hole
h_2 = back of top hole)

Consortini, P., *Volterra Nell'Antichita*, Volterra, 1940, p. 198 (photograph).
CIL, XI, 1, 1749.

23

24

2018G. Spherical Variant
Bologna, Italy
Civic Museum, portico

White stone - Nine hour lines (2,3,4,5,6,7,8,9,10), the equinox, and the summer solstice are preserved. The winter solstice is obliterated. The lightly incised curves are sketchy and approximate. The back is squared; the top, flat.

H = 385 mm
W = 425 mm
D = 260 mm

te = 227 mm
es = 156 mm
sb = 14 mm

(I to II)
e = 69 mm
(VI to VII)
s = 106 mm

$\theta = 130°$

For $\varepsilon = 24°$, $r_s = 186$ mm

$\phi_s = 45°$, for $(\text{I to II})_e = 69$ mm

$\phi_s = 44°13'$, for $(\text{II to III})_e = 68$ mm

$\phi_s = 47°30'$, for $(\text{III to IV})_e = 72$ mm

Unpublished

2019G. Spherical Variant
Pompeii, Italy
Granario

Plate 25
Page 221

White marble - The spherical shadow-receiving surface of this dial has been hollowed out of a bowl-shaped stone complete with handle preserved on the right side. A second larger protrusion serves as the dial's base. When set up on this base the edge of the spherical hollow leans forward 26° from normal. Seven hour lines (4,5,6,7,8,9,10) extend between the two solstices and are crossed by the equinox.

PH = 550 mm
PW = 530 mm
r = 235 mm
tw = 80 mm
we = 250 mm
es = 265 mm
sb = 5 mm
(VI to VII)
w = 46.5 mm
e = 99 mm
s = 150 mm
θ = 116°

For $\varepsilon = 24°$, and $(VI\ to\ VII)_e = 99$ mm, $r_w = 298$ mm, $\phi_w = 50°$, $r_s = 316$ mm and $\phi_s = 53°$

Dial 2001G from Altino also has bowl-shaped sides and back.

Unpublished

2020. Spherical Variant
Madrid, Spain
National Archeological Museum, no. 33.185

A round hole (190.5 mm in diameter) bordered by a groove pierces the top of the dial. Some lead remains in one of four depressions in the groove.

H = 850.9 mm
W = 723.9 mm
D_{top} = 596.9 mm

Marble - Paris reports existence of eleven hour lines and three day curves on the spherical shadow-receiving surface. Two well-formed lion feet decorate the base of the dial and two rosettes decorate the top corners. A relief margin emphasizes the front circular opening.

Found at Belo, Cadiz (ancient Βαίλων) in the kitchen of the "House of the Sundial." Paris somehow determines that the dial was made for latitude 41°30' and concludes that it came from Rome, probably originally in a public place.

The decorative details on this dial are quite similar to those on 2023G although of much finer quality. It is possible that Paris measured α = 48°30'.

Paris, P., and Bonsor, G., *Fouilles de Belo*, Paris, 1923, p. 142, fig. 48 (photograph); p. 166, fig. 64 (drawing); p. 167, fig. 65 (drawing, profile).
Cf. *Museo Arqueologico Nacional*, Madrid, 1965, p. 59, no. 59.

2021G. Spherical Variant
Vatican City
Vatican Museum, Cortile Otagono

Travertine marble - The six preserved hour lines (4,5,6,7,8,9) on this dial are arranged symmetrically about the meridian. Only parts of the equinox and summer solstice remain; the winter solstice is entirely broken away. The base was flanked by two feet with nobby widely-spaced toes; one of the pair is missing.

H = 370 mm
W = 369 mm
D_{bottom} = 227 mm
es = 134 mm
(VI to VII)
e = 63 mm
s = 93 mm
θ = 126.5°

For $\varepsilon = 24°$, r_s = 160 mm
ϕ_s = 41°13'

Note that the lion foot on this dial is very similar to the feet on dial 2017G from Volterra.

Cf. Antonini, dial 2 (drawing).

2022. Spherical Variant
Rome, Italy
Palatine Antiquarium

The same photograph of this restored roofed spherical dial has been published in the two sources listed below. Unfortunately a search of the storage rooms of Palatine finds did not produce the dial in the summer of 1971.

Note similarity to 2003G from Aquileia.

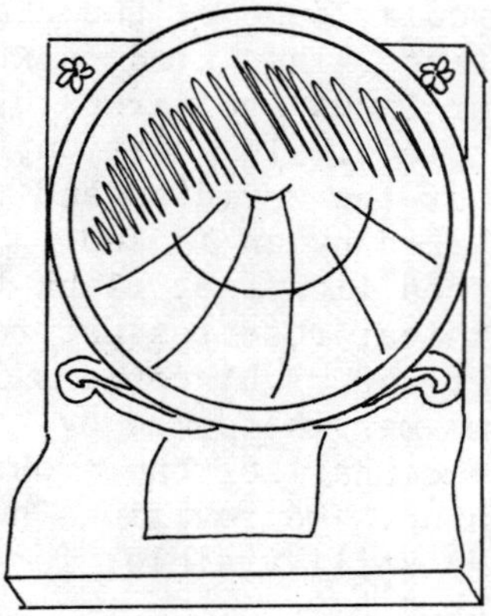

Cf. Gabba, L. and Jacono, L. "Meridiana," *Enciclopedia Italiana*, Rome, 1935-43, p. 903 (photograph).
Portaluppi, Piero, *Gnomonica Atellana*, Milan, 1968, p. 19 (photograph).

2023G. Spherical Variant
Berlin, East Germany
Pergamon Museum, Antiken Sammlung, no. SK1049

$H = 484$ mm
$W = 399$ mm (base)
$W = 410$ mm (top)
$D_{base} = 229$ mm
$D_{top} = 480$ mm
$d = 332$ mm
$tw = 243$ mm
$we = 141$ mm
$es = 141.5$ mm
(V to VI)
$w = 31$ mm
$e = 68.2$ mm
$s = 94$ mm
$\theta = 131.4°$

A 15.5 mm square depression with a narrow rectangular appendage is preserved in the top surface of the dial. It is pierced by two small holes 4 mm in diameter toward the front and, in the center, a large broken hole. Both large and small holes are also visible in the spherical surface.

White marble - Eleven hour lines and three day curves are preserved. The summer solstice coincides with the front circular edge of the dial for some distance. The third, sixth, and ninth hour lines are marked by a short crosspiece. In addition, the juncture of the meridian and the equinox is marked by an X. Two squat, four-toed lion feet flank the base. Two six-leaf rosettes decorate the upper left and right corners in front. A 16 mm margin formed by an engraved line emphasizes the front circular opening. The revisions noted by Drecker are still visible, for example in lines 7,8,9.

For $r = 168$ mm, $(V \text{ to } VI)_e = 68.2$ mm, and $\varepsilon = 24°$, $\phi = 39°$.

Drecker, pp. 25-34 and fig. 41 (photograph).
Kekule, W. S., *Beschreibung der Antiken Skulpturen*, Berlin, 1891, p. 417, no. 1049.
Woepcke, F., *Disquisitiones Archaeologico-mathematicae circa Solaria Veterum*, Berlin, 1847.
Cf. Diels, pp. 167-68 and plate 11 (photograph).
Gatty, p. 35 (drawing).

Conical Dials

Measured Quantities

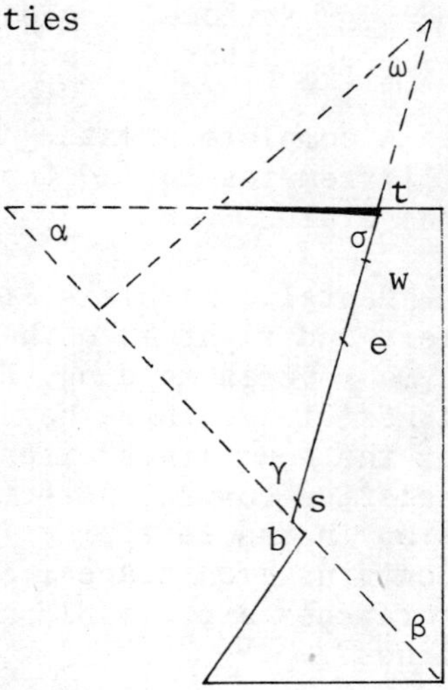

H = height (P = partial)
W = width
D = depth

tw = distance along meridian from top edge of dial to winter solstice curve
gw = distance along meridian from lower edge of gnomon to winter solstice curve
we = distance along meridian from winter solstice curve to equinox curve
es = distance along meridian from equinox curve to summer solstice curve
sb = distance along meridian from summer solstice curve to bottom of dial face

(VI to VII)
w = distance along the winter solstice curve from the sixth to the seventh hour points
e = distance along the equinox curve from the sixth to the seventh hour points
s = distance along the summer solstice curve from the sixth to the seventh hour points

α = acute angle between oblique front of dial and horizontal
β = acute angle between oblique front of dial and vertical
γ = angle between oblique front of dial and dial face
σ = angle between horizontal and dial face

Computed Quantities

g = gnomon length
ω = angle between axis and generator of conical surface
ε = obliquity of ecliptic
ϕ_s = latitude computed from measurements along summer solstice
ϕ_w = latitude computed from measurements along winter solstice

3001G. Conical
Athens, Greece
National Archeological Museum, study collection, no. 3158

Plate 26

A complete bronze gnomon in lead fill remains in the top surface of the dial.

Pentelic marble - Stylized legs at left and right have the profile of a cyma reversa molding. Eleven hour lines divide three day curves. None of the hour lines extend beyond the solstice curves. A rectangular depression in the left side (54 mm x 44 mm) contains iron traces. Some light scratches are visible near the hour lines.

Found in the ruins of the theater of Dionysos in Athens.

$H = 220$ mm
$W = 242$ mm
$D_{top} = 181$ mm
$D_{base} = 155$ mm
$gw = 16$ mm
$we = 31.5$ mm
$es = 62.0$ mm
(VI to VII)
$w = 10$ mm
$e = 18$ mm
$s = 32$ mm
$\gamma = 58.5°$
$\beta = 38°$

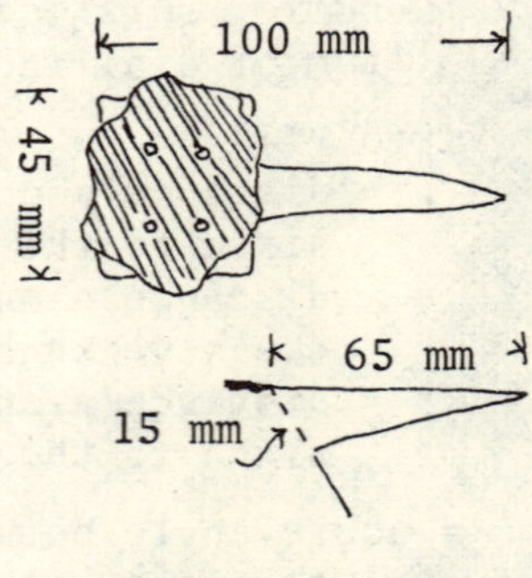

gnomon

For $\varepsilon = 24°$, $\omega = 36°30'$
$g = 59.37$ mm
$\phi = 32°12'$
$\phi_W = 42°$
$\phi_S = 34°30'$

Other conical dials with preserved gnomon include 3023G, 3024G, and 3064G.

von Basserman-Jordan, E., *Uhren*, Berlin, 1922, p. 18, fig. 7 (photograph).
Heydemann, H., *Die Antiken Marmor-Bildwerke zu Athen*, Berlin, 1874, p. 285, no. 776.
von Sybel, Ludwig, *Katalog der Sculpturen zu Athen*, Marburg, 1881, p. 128, no. 903.

25

26

3002G. Conical
Athens, Greece
Museum of the Stoa of Attalos, no. A1769

H = 300 mm
W = 240 mm
D = 235 mm

along third hour line
we = 45 mm
es = 90 mm

(III to IV)
w = 16 mm
e = 28 mm

A gnomon hole 34 mm deep and black stained is visible in the top surface.

Pentelic marble - Parts of four hour lines (2,3,4,5) and two day curves (winter solstice and equinox) are preserved. The hour lines do not extend beyond the winter solstice, and day curves are parallel. A portion of the leg support in high relief is visible at lower left. A trace of leaf pattern remains on left side.

Found in a marble dump in the Greek Agora at Athens, April 12, 1951.

Marble used to construct this dial is similar to that used for at least two other dials found in the Agora: 3012G and 3013G.

Unpublished

3003G. Conical
Athens, Greece
National Archeological Museum, no. 3220

White marble - The top has been sheared off this unique dial. Nine hour lines extend from the sheared top to the damaged lower edge. The cylindrical back looks like part of a column without flutes. Lion legs and feet flank the dial surface. A hole 24 mm in diameter intersects the meridian near the lower edge and penetrates through to the bottom.

H = 217 mm
W = 433 mm
PD = 180 mm

σ = 56.5°

Unpublished

***

3004G. Conical
Athens, Greece
Epigraphical Museum, no. 9818

Plate 27
Page 228

White marble with streaks of gray - Only one day curve crosses the seven preserved hour lines (1,2,3,4,5,6,7) 6 mm from the lower edge. Ferrous stains are evident in thumbprint-shaped remains of mount hole at the center of the bottom front. A Greek inscription on the front lip and also on the base between the remains of relief legs reads:

Σεβαστῷ
᾿Ανθεσ/
τῆ[ριο]s

H = 298 mm
W = 298 mm
PD = 210 mm

tb = 140 mm

(VI to VII)
c = 40 mm

σ = 73.5°
β = 38°
γ = 55°

CIA, III, 2, 3857 (drawing).

3005G. Conical
Athens, Greece
National Archeological Museum, no. 3157

The preserved portion of the gnomon groove is 33 mm long. Traces of lead fill are visible.

H = 287 mm
W = 340 mm
PD = 260 mm

White marble - Eight hour lines (4,5,6,7,8,9,10,11) extend from the winter solstice across the equinox and the summer solstice to the lower edge. Only the meridian continues above the winter solstice to the gnomon hole. Two legless lion feet flank the base.

tw = 60 mm
we = 48.5 mm
es = 68.5 mm
sb = 3 mm

(VI to VII)
w = 22 mm
e = 32 mm
s = 45.2 mm

Found during the excavation of the theater of Dionysos, Athens.

$\sigma = 61°$
$\gamma = 68°$
$\beta = 39°$

For $\varepsilon = 24°$, $\omega = 21°$, g = 131mm,
and $\phi = 39°24'$, with $\phi_W = 34°6'$
and $\phi_S = 30°48'$

A conical dial with similar legless feet is now preserved in the Basmahane Archeological Museum, Izmir, Turkey (3060G).

Heydemann, H., *Die Antiken Marmor-Bildwerke*, Berlin, 1874, p. 102, no. 268.
von Sybel, *Katalog der Sculpturen zu Athen*, Marburg, 1881, p. 128, no. 931.

3006G. Conical
Berlin, East Germany
Pergamon Museum, Antiken Sammlung, no. 1048

The gnomon hole is 20 mm deep.

Pentelic marble - Only two engraved hour lines are visible. Eleven hour lines extend beyond the winter solstice but do not continue to the gnomon hole. two fine profiles of Athena (left) and Dionysos (right) in low relief decorate the sides. A head of Helios is carved into the center of the base flanked by two stylized legs. A geometric design decorates the back ("the ornament of the Patera") and small leaves have been engraved on the sides in the angle between the top and front edges.

H = 212 mm
W = 230 mm
D_{top} = 190 mm

tw = 24 mm
we = 25 mm
eb = 67.5 mm

(VI to VII)
w = 6.5 mm
e = 13 mm
s = 33 mm

$\sigma = 78.2°$
$\alpha = 54°$

The winter solstice curve is divided into arcs of the following lengths: 9, 10, 9, 7.5, 5.5, 6.5, 6.5, 7, 7.5, 9, 7 mm. The equinox curve is divided into arcs of the following lengths: 15, 15.5, 16, 15, 13, 12, 13, 14, 13.5, 17.5, 15, 13 mm.

For $\varepsilon = 24°$, $\omega = 45°30'$, g = 39.5 mm, and $\phi = 30°$, with $\phi_W = 39°54'$
and $\phi_S = 39°42'$

A similar corner leaf design is visible on 3027G.

AZ 38 (1880): 37.
Kekule, W. S., *Beschreibung der Antiken Skulpturen*, Berlin, 1891, p. 417, no. 1048 (drawing).
Cf. Diels, p. 175 and plate 14 (drawing)..
Drecker, p. 37 and figure 60 (photograph).
Gatty, p. 33.

3007. Conical
Leiden, The Netherlands
Rijksmuseum van Oudheden, no. RO III 65

A large gnomon hole in the top surface of the dial has the shape of a triangular prism, 40 mm long, 30 mm wide, and 23 mm deep.

H = 230 mm
W = 260 mm
D = 190 mm

tw = 36 mm
we = 34 mm
es = 58 mm
sb = 6 mm

$\gamma = 60°$
$\beta = 37°$

White marble - Three engraved day curves are evenly divided by eleven hour lines. The noon line extends from the gnomon groove to the lower edge. Short half hour marks divide the summer solstice curve but do not extend to the equinox. Two schematized feet flank the base.

Found in Athens near the ancient Academy and placed in the Leiden Museum in 1826.

This dial is characterized as Greek by a curator of the museum. The same curator has calculated the radius of the front edge to be 115 mm with the circular arc of the edge corresponding to 215°. He also found the top horizontal section to be part of an elipse with long axis 534 mm and short axis 235 mm.

For $\varepsilon = 24°$, $\omega = 30°18'$.

A second conical dial with half hour markings was found at Megara (3020G).

Measurements taken in May, 1958, by P. van der Star, Assistant Curator of the Rijksmuseum voor de Geschiedenis der Natuurwetenschappen.

Cf. Drecker, p. 37.
Gatty, pp. 31-32 (drawing).

3008G. Conical
Plate 28
Page 228

Athens, Greece
In situ east of the monument of Thrasyllos below the Acropolis above the theater of Dionysos

Traces of lead and ferrous stains remain in a large gnomon hole above the meridian.

White marble - Eleven hour lines extend across the equinox between the solstices. All lines and curves are 9 mm wide and 5 mm deep. The rough, pitted surface in these grooves is discolored by black granules. Two stylized legs support the dial face.

H = 860 mm
W = 1230 mm
D = 820 mm

tw = 122 mm
we = 137 mm
es = 342 mm
sb = 9 mm

(VI to VII)
w = 49 mm
e = 86 mm
s = 171 mm

$\sigma = 82°$
$\gamma = 42°$
$\beta = 37.5°$

For $\varepsilon = 24°$, $\omega = 43°30'$
$g = 240.6$ mm
$\phi = 38°36'$
$\phi_W = 33°$
$\phi_S = 25°42'$

LeRoy, *Ruines des Plus Beaux Monuments de la Grèce*, vol. 2, plate 2, Paris, 1770.
Cf. Gatty, p. 34.
Stuart, J. and Revette, N., *The Antiquities of Athens*, vol. 2, London, 1786, p. 29 vignette (drawings), p. 46, plate 1.

27

28

3009G. Conical
Athens, Greece
Museum of the Stoa of Attalos, no. ST589

Pentelic marble - The back and right side are preserved on this otherwise badly damaged fragment. Only three hour lines (7,8,9) extend from the winter solstice to the damaged lower edge. The two preserved day curves (winter solstice and equinox) appear to be less deeply engraved than the hour lines.

Recovered from a pre-war marble pile in the Greek Agora of Athens, May, 1953.

H = 310 mm
W = 310 mm
D = 90 mm

along eighth hour line
we = 55 mm
along ninth hour line
we = 57.5 mm

(II to III)
w = 18 mm
e = 31.5 mm
s = 56 mm

For other dials made from similar marble and found in the Agora, see: 3014G, 3002G, 3012G, and 3013G.

Unpublished

3010G. Conical
Athens, Greece
Museum of the Stoa of Attalos, no. ST147

H = 159 mm
W = 130 mm
D = 120 mm

gw = 22.5 mm
we = 22.5 mm
eb = 44 mm

(VI to VII)
w = 9.5 mm
e = 13.5 mm
s = 23 mm

$\sigma = 66°$
$\gamma = 61°$
$\beta = 38.5°$

A lead shell remains in this 31 mm long wedge-shaped gnomon hole.

Pentelic marble - Ten hour lines are preserved on this small, worn example (2,3,4,5,6,7,8,9,10,11). All hour lines extend beyond a winter solstice which is parallel to the engraved equinox and the lower edge. Day curves appear to be less deeply engraved than hour lines. Two stylized legs flank the base.

Found in well at 6/NE in the Greek Agora at Athens, May 22, 1936.

For $\varepsilon = 24°$, $\omega = 36°$
$g = 44$ mm
$\phi = 34°36'$
$\phi_W = 13°36'$
$\phi_S = 28°$

This dial is considerably smaller than the average conical dial.

Unpublished

3011G. Conical
Athens, Greece
Museum of the Stoa of Attalos, no. ST780

The gnomon hole in the top surface is 48 mm long, 8 mm wide, and 38 mm deep.

H = 150 mm
W = 320 mm
D = 220 mm

gw = 50 mm
we = 44 mm
es = 74 mm

(VI to VII)
w = 19 mm
e = 28 mm
s = 47 mm

$\sigma = 72°$
$\gamma = 51.5°$
$\beta = 37.5°$

Pentelic marble - Three lightly engraved parallel day curves cross eleven heavy hour lines. All hour lines extend beyond the winter solstice toward the gnomon hole. Two lion feet decorate the sides and the base. The side decoration also includes a rosette, calyx, and acanthus stalk, all of which spring from the calf of the lion's leg.

Found August 4, 1970, in the upper destruction fill of a Roman house in the Greek Agora at Athens.

For $\varepsilon = 24°$, $\omega = 30°$
$g = 101$ mm
$\phi = 36°24'$
$\phi_w = 27°30'$
$\phi_s = 40°48'$

Dials with similar side reliefs include 3016G and 3018G from Corinth, and 3027G from Delos.

The marble used to construct this dial is similar to that used to construct 3012G and 3013G, also found in the Greek Agora at Athens.

Unpublished

3012G. Conical
Athens, Greece
Museum of the Stoa of Attalos, no. ST148

H = 170 mm
W = 155 mm
D = 112 mm

gw = 30 mm
we = 30 mm
es = 55 mm

(V to VI)
w = 11 mm
e = 20 mm

$\sigma = 70°$
$\gamma = 54°$

The preserved portion of the gnomon hole is 31 mm deep.

Pentelic marble - The two preserved day curves appear to have been scratched on after the hour lines were deeply engraved. Some of the six remaining hour lines (left half and meridian) extend beyond the winter solstice. A series of steps seems to have decorated the base.

Found June 9, 1939, in the midst of late Roman fill in the Greek Agora at Athens.

For $\varepsilon = 24°$, $\omega = 33°30'$
$g = 66$ mm
$\phi = 40°48'$
$\phi_w = 44°30'$

The marble used to construct this dial is similar to that used for 3002G, 3011G, and 3013G, and also found in the Agora in Athens.

Unpublished

3013G. Conical
Athens, Greece
Museum of the Stoa of Attalos, no. ST209

A distinctively shaped gnomon hole, 31 mm deep, is visible in the top surface

Pentelic marble - Six hour lines (2,3,4,5,6,7) extend between two solstices which are parallel to each other and to the equinox. Part of a leg base support is preserved at the lower left.

Found in a marble pile in the Greek Agora at Athens, June 3, 1937.

H = 220 mm
W = 165 mm
D = 114 mm

gw = 32 mm
we = 32.5 mm
es = 50.5 mm

(II to III)
w = 11 mm
(III to IV)
e = 18.5 mm
(IV to V)
s = 31 mm

For $\varepsilon = 24°$, $\omega = 26°$
$g = 72$ mm
$\phi = 36°12'$
$\phi_W = 43°24'$
$\phi_S = 46°36'$

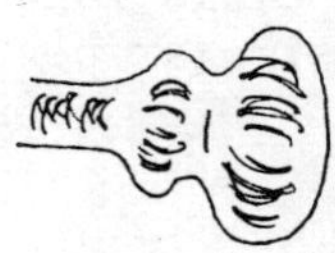

gnomon hole

The marble used to construct this dial is similar to that used for 3002G, 3011G, and 3012G, all from the Athenian Agora.

Unpublished

3014G. Conical
Athens, Greece
Museum of the Stoa of Attalos, no. A1870

Pentelic marble - The fine lines preserved on this fragment include four hour lines (3,4,5,6) and two day curves (equinox and summer solstice). Part of a relief leg support is preserved at the lower left front.

Found in marble pile north of the temple of Ares in the Greek Agora at Athens, on July 2, 1951.

Other dials found in the Agora made from similar pentelic marble include 3002G, 3009G, 3012G, 3013G.

H = 120 mm
W = 165 mm
D = 180 mm
along fifth hour line
es = 62.5 mm
along sixth hour line
es = 63 mm
(IV to V)
e = 25 mm
s = 40 mm
$\gamma = 49°$
$\alpha = 53.5°$

Unpublished

3015G. Conical
Athens, Greece
National Archeological Museum, no. 3156

A rounded rectangular gnomon hole (34 mm x 38 mm) is preserved in the top surface of the dial.

White marble - Eleven hour lines extend between two solstices and are crossed by the equinox. Two duckpin shaped lion legs and feet decorate the base.

H = 502 mm
W = 430 mm
D = 295 mm

gW = 15.5 mm
we = 50.5 mm
es = 116 mm

(VI to VII)
w = 19 mm
e = 32 mm
s = 63 mm

$\sigma = 80°$
$\gamma = 43.5°$
$\beta = 33°$

For $\varepsilon = 24°$, $\omega = 41°28'$
$g = 92.8$ mm
$\phi = 39°11'$
$\phi_W = 31°$
$\phi_S = 34°$

Rayet's calculated parameters based on measurements taken by the director of the French School at Athens include $\alpha = 52°$, $\beta = 38°$, $\gamma = 47°50'$, $\omega = 42°10'$. He concludes that the conic section at the top edge is a hyperbola with real axis: 2 x 275.5 mm, and imaginary axis: 2 x 133.4 mm.

Rayet, pp. 72-74.

3016G. Conical
Ancient Corinth, Greece
Museum, no. A886

A rectangular gnomon hole is preserved in the top surface.

Small crystal white marble - Parts of eight hour lines (4,5,6,7,8,9,10, 11) remain on this example. Only the meridian continues beyond the winter solstice to the gnomon hole. The equinox appears to be parallel to the winter solstice; only a trace of summer solstice is preserved along the damaged lower edge. The winged calf of an animal leg and a wheel with eight spokes are visible in the relief design on the right side.

Found near the Odeum in the wall to the west of the dig on May 4, 1927.

H = 266 mm
W = 262 mm
PD = 125 mm

tw = 42.5 mm
we = 42.5 mm
es = 68+ mm

(VI to VII)
w = 16 mm
e = 25.5 mm
(V to VI)
s = 40 mm

$\sigma = 63°$

For $\varepsilon = 24°$, $\omega = 27°18'$, $g = 95.4$ mm, and $\phi = 37°42'$, with $\phi_W = 38°12'$ and $\phi_S = 35°36'$.

Dials with similar side reliefs include 3018G, also from Corinth, 3027G from Delos, and 3011G from Athens.

Unpublished

3017G. Conical
Ancient Corinth, Greece
Museum, no. A930

Pentelic marble - Eleven hour lines extend from the broken top to the lower edge. No day curves cross the hour lines. Two duckpin-shaped lion legs and feet flank the base. The back is cylindrical.

Found in 1928 during the excavations of Professor L. T. Shear.

For other conical dials with hour lines only see 3003G, 3018G, 3044G, 3066G, 3088, 3091, 3099, 3103.

H = 335 mm
W = 264 mm
PD = 155 mm
tb = 165+ mm
(VI to VII)
b = 33.5 mm
$\sigma = 51°$
$\gamma = 57°$
$\beta = 18°$

Unpublished

***

3018G. Conical
Ancient Corinth, Greece
Museum, no. A924

Plate 29
Page 241

Part of a 50 mm deep gnomon hole remains in the top surface.

Fine grained white marble - Eight preserved hour lines extend from various places along the top edge (4,5,6,7, 8,9,10,11). No day curves are visible. A mount hole remains in the lower surface below the rosette in the center of the base which consists of two rows of eight petals each. The upper portions of two lion feet flank the rosette. A relief design on the left and right sides consists of a spiral and a sprouting calyx.

H = 340 mm
W = 354 mm
PD = 215 mm
$\sigma = 69°$

Unpublished

3019G. Conical
Isthmia, Greece
Excavations of Dr. Paul Clement

A trace of lead remains in the bottom of the rectangular gnomon hole 38 mm x 15 mm.

Pentelic marble - Eleven hour lines and two engraved day curves are visible on this beautifully preserved example. The summer solstice apparently coincides with the lower edge. Some hour lines extend beyond the winter solstice, but only the meridian continues all the way to the gnomon hole. Two relief legs with cyma recta profiles flank the base.

Part of late Roman fill from Isthmia, Greece.

H = 459 mm
W = 466 mm
D_{bottom} = 356 mm
D_{top} = 328 mm
gw = 52.5 mm
we = 57.5 mm
eb = 116 mm
(VI to VII)
w = 22 mm
e = 37 mm
l = 65 mm
$\sigma = 69°$
$\gamma = 58°$
$\beta = 38°$

For $\varepsilon = 24°$, $\omega = 37°20'$
$g = 115$ mm
$\phi = 40°30'$
$\phi_W = 35°24'$
$\phi_S = 29°48'$

For gw = 57.5 mm, $\phi = 38°6'$

Unpublished

3020G. Conical
Megara, Greece
Study collection in gymnasium, no. 42

Plate 30
Page 241

The gnomon hole preserved in the top surface of the dial is 32 mm long, 10 mm wide, and 44 mm deep.

Pentelic marble -- All eleven hour lines on this dial extend from the winter solstice across the equinox to the lower edge. In addition, half hour lines extend between the equinox and the lower edge. The first visible half hour line (between the first and the second hour line) extends from the winter solstice to the lower edge. The lower edge is emphasized by a deep groove cut into the front surface. The remains of two animal legs flank the base. A short groove in the lower front marks the center of the base.

H = 400 mm
W = 400 mm
D = 200 mm

tw = 55.5 mm
we = 50.5 mm
eb = 94 mm

(VI to VII)
w = 22.2 mm
e = 35.5 mm
s = 56 mm

$\sigma = 70°$
$\gamma = 53°$
$\alpha = 36.5°$

For $\varepsilon = 24°$, $\omega = 34°$
$g = 116.4$ mm
$\phi = 41°$
$\phi_W = 32°45'$
$\phi_S = 19°48'$

A conical dial (3007) now in the Rijksmuseum, Leiden, also has half hour markings.

Unpublished

3021G. Conical
Oropus-Amphiareion, Greece
Museum

Plate 31

A rectangular gnomon hole, 38 mm long, 20 mm wide, and 35 mm deep remains in the center of the top surface of the dial.

Cream marble - All eleven engraved hour lines continue to the gnomon hole. Three parallel day curves, including a badly worn winter solstice, cross these hour lines. Part of an inscription is preserved in the center of the base:

Ιασὼν Ιασονος
ʼΑζηṿ[ιεύs--]ηι

A square mount hole is preserved in the center of the bottom.

For ε = 24°, ω = 33°30', g = 92 mm, and ϕ = 40°, with ϕ_W = 21° and ϕ_S = 31°.

H = 245 mm
W = 348 mm
D_{max} = 177 mm

tw = 44.5 mm
we = 40 mm
es = 73.5 mm
sb = 6 mm

(VI to VII)
w = 19.1 mm
e = 27.8 mm
s = 46.3 mm

σ = 70°
γ = 56.2°
β = 37°

Unpublished

***

3022. Conical
Olympia, Greece
Archeological Museum, storeyard

Trace of lead remains in gnomon hole.

Pentelic marble - Three engraved day curves and ten hour lines are visible. The third, sixth, and ninth hour lines are marked: X, O, X. Two six-petalled rosettes flank a crescent moon on the oblique front surface. A half-cylinder molding decorates the base.

Moore, R., and Kazantzakis, N., *Reflections on Greece*, New York, 1971 (photograph).

29

30

31

3023G. Conical — Plate 32, Page 245
Piraeus, Greece
Archeological Museum, Meletopoulos Collection

A complete bronze gnomon remains in lead fill at the top of the dial.

White marble - An 18.7 mm wide relief lip borders the front circular section. Each double concave stylized leg is surmounted by an engraved triangle. Eleven hour lines equally divide the three parallel day curves. The meridian continues beyond the winter solstice to the gnomon. A dark red color in the meridian line and day curves. Black color is visible in the morning hour lines.

H = 220 mm
W = 187 mm
D_{top} = 165 mm
D_{base} = 125 mm
gw = 15 mm
we = 22.5 mm
es = 55 mm
sb = 3 mm

(VI to VII)
w = 10 mm
e = 16.5 mm
s = 26.5 mm

$\gamma = 71°$
$\beta = 37°$

Paris reports seeing construction-related markings on the side and top. I found only a trace of the meridian engraved on the oblique front.

For $\varepsilon = 24°$, $\omega = 43°20'$, g = 46.8 mm, and $\phi = 35°12'$, with $\phi_W = 26°$, and $\phi_S = 13°24'$
Paris suggests possible dialler's error.

Other conical dials with gnomon preserved include 3024G, 3048, 3064G, and 3001G.

Construction markings are also visible on 3025G, 3028G, 3067G.

Paris, J., "Gnomonica, Un Cadran Solaire Ancien du Pirée," *Bulletin Bibliographique et Pedagogique du Musée Belge Revue de Philologie Classique* 18(1914):121-30, figure 1 (drawing).

3024G. Conical — Frontispiece
Delos, Greece
Archeological Museum, study collection
no. B3652/11023

A broken bronze gnomon remains in the lead filled hole in the top surface of the dial.

White marble - Narrow stylized legs, flanking the base, have a cyma reversa profile. All eleven hour lines continue beyond the winter solstice; the meridian continues to the gnomon. None of the hour lines continue beyond the summer solstice. There are traces of a dark red color in the three day curves as well as similar red color in the hour lines. An 11 mm band of red color emphasizes the circular section at the front of the dial.

Found in 1910 northwest of the bank of the Sacred Lake in Delos.

H = 230 mm
W = 227 mm
D_{top} = 138 mm
D_{base} = 135 mm
gw = 14 mm
we = 31.5 mm
es = 55.5 mm
sb = 7.5 mm
(VI to VII)
w = 11 mm
e = 19 mm
s = 33.5 mm
$\gamma = 43.5°$
$\beta = 39°$

For $\varepsilon = 24°$, $\omega = 32°$
$g = 65.4$ mm
$\phi = 36°18'$
$\phi_w = 41°45'$
$\phi_s = 43°24'$

Other conical dials with original gnomon in place include 3001G, 3048, 3023G, 3064G.

Red color is also visible in the day curves on 3031G, 3023G, 3092, 1023G, 1024G, 1026G, 1002G.

Deonna, p. 190, figures 220-23 (drawings), plate 540 (photograph).

3025G. Conical
Delos, Greece
Museum, portico

Plate 33

Semicircular gnomon hole is 27.5 mm deep.

White marble - The face of this dial is so badly worn that the position of the seasonal indicators cannot be determined. However, parts of eleven hour lines can be distinguished. A series of lines related to the construction of this dial are preserved on its top surface. On the base, the area between cyma reversa legs is inscribed in large Greek letters.

ΙΗΝΩ
ΝΩΝΙ
[Η]ΡΩΙ

H = 198 mm
W = 199 mm
D_{max} = 135 mm
tb = 112.5 mm
(VI to VII)
b = 28.5 mm
σ = 58°
γ = 68°
β = 37.5°

It seems certain that with the exception of the first letter of the third line, the inscription is complete as transcribed. The projecting flanking legs would have made continuation difficult.

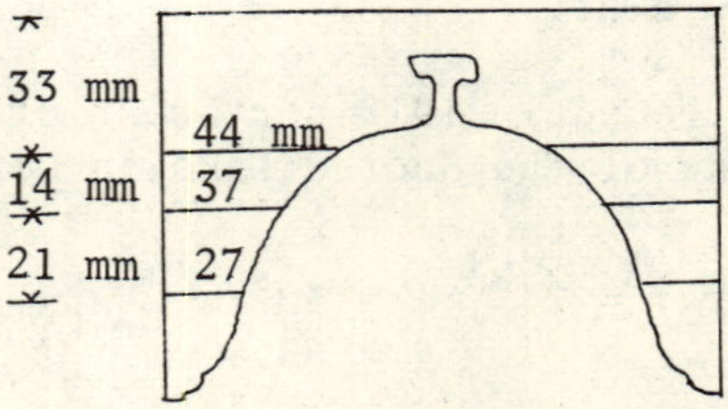

Deonna, p. 190, plate 537 (photograph).
Cf. Preuner, E., "Aus alten Papieren II," Ath. Mitt. 49 (1924): 143.

32

33

3026G. Conical
Delos, Greece
Museum, storage room

A half-cylindrical gnomon hole and connecting groove are preserved in the top surface.

White marble -- Eleven hour lines extend from the area near the gnomon hole across winter solstice, equinox, and summer solstice to the lower edge. Parts of two animal legs flank the base. The left and right sides are decorated with reliefs depicting flowering stalks and tendrils.

H = 400 mm
W = 468 mm
D_{max} = 280 mm

gw = 58.5 mm
we = 58.5 mm
es = 101 mm

(VI to VII)
w = 24 mm
e = 37.5
s = 65 mm

$\sigma = 64°$
$\gamma = 64°$

For $\varepsilon = 24°$, $\omega = 31°$
g = 130 mm
$\phi = 39°36'$
$\phi_W = 33°24'$
$\phi_S = 42°42'$

Dials with similarly ornamented sides and lion feet include 3079G, 3071G, 3064G.

Deonna, p. 190, fig. 215 (drawing), pl. 535 (photograph).

3027G. Conical — Plate 34
Delos, Greece — Page 255
Museum, storage room, no. B 1046-6047

The gnomon hole in the top surface is square.

Pentelic marble - Eight distinctly engraved hour lines (1,2,3,4,5,6,7,8) and parts of three day curves are preserved on this dial surface. The hour lines do not extend beyond the solstices. A wide lip follows the curve of the lower edge. The lion paws which flank the base continue onto the side reliefs which also include a rosette base and a calyx with petals and tentril.

Found north of the agora of Theophrastos in Delos.

H = 318 mm
W = 310 mm
D_{top} = 233 mm
D_{bottom} = 220 mm
tw = 33.5 mm
we = 23 mm
es = 81 mm
sb = 3 mm

(II to III)
w = 12 mm
e = 20.5 mm
s = 45 mm

$\sigma = 90°$
$\gamma = 35°$
$\alpha = 52°$

For $\varepsilon = 24°$, $\omega = 51°20'$
g = 78 mm
$\phi = 25°12'$
$\phi_W = 17°12'$
$\phi_S = 5°51'$

Dials with similar side reliefs have been found at Corinth (3018G, 3016G) and Athens (3011G).

Deonna, p. 190, figs. 216 and 217 (drawings), plate 536 (photograph).

3028G. Conical
Delos, Greece
Museum, portico, no. B 115-6243

The base of a bronze gnomon in lead fill remains in the top of the dial above the meridian.

White marble - Lightly engraved guide lines for the stone cutter remain in the oblique front surface of the dial. They include three lines which follow the curve of the lower edge and a continuation of the meridian. On the shadow-receiving surface, eleven hour lines extend from the winter solstice across the equinox and the summer solstice to the lower edge. The meridian line continues to the gnomon. Two legs with cyma reversa profile flank the base.

Found northwest of the oikos of Dionysos on Delos.

H = 211 mm
W = 230 mm
D = 145 mm

gw = 26 mm
we = 28.5 mm
es = 45.5 mm
sb = 6.5 mm

(VI to VII)
w = 10 mm
e = 17.2 mm
s = 28 mm

$\sigma = 72.5°$
$\gamma = 51.5°$
$\alpha = 52.5°$

For $\varepsilon = 24°$, $\omega = 27°20'$
$g = 64.10$ mm
$\phi = 38°$
$\phi_W = 44°19'$
$\phi_S = 65°18'$

Deonna, W., p. 190, figs. 218 and 219 (drawings), pl. 538 (photograph).

3029G. Conical
Delos, Greece
Museum, portico, no. 5394

White marble - The base and left half are broken away from this example. However, seven hour lines remain (5,6, 7,8,9,10,11) extending from the winter solstice across the equinox to the summer solstice and beyond.

$H = 131$ mm
$W_{max} = 166$ mm
$D_{max} = 118.5$ mm
tw = 34 mm
we = 34 mm
es = 55.5 mm
sb = 6 mm
(VI to VII)
w = 12.5 mm
e = 21 mm
s = 34 mm
$\gamma = 48°$
$\alpha = 56.5°$

For $\varepsilon = 24°$, $\omega = 28°20'$
$g = 77$ mm
$\phi = 38°68'$
$\phi_W = 42°$
$\phi_S = 39°$

Deonna, p. 191, fig. 225 (drawings).

3030G. Conical
Delos, Greece
Museum, portico, no. B1001-5941

A rectangular gnomon hole and related groove remain in the top surface of the dial.

White marble - Preserved portions of ten hour lines (2,3,4,5,6,7,8,9, 10,11) extend from the winter solstice across the equinox to the summer solstice. The meridian extends from the gnomon groove to the lower edge. A trace of squarish leg is visible in the lower right front. A round mount hole is preserved in the bottom surface.

Found north of the agora of Theophrastos on Delos.

H = 280 mm
W = 282 mm
D_{max} = 170 mm

gw = 58 mm
we = 40.5 mm
es = 55 mm
sb = 16 mm

(VI to VII)
w = 19 mm
e = 27 mm
s = 38 mm

$\sigma = 55.5°$
$\gamma = 70°$
$\beta = 35.5°$

The agora was set up by the Magistrate Theophrastos in 126/25 B.C.

For $\varepsilon = 24°$, $\omega = 19°$
$g = 117$ mm
$\phi = 37°34'$
$\phi_W = 33°24'$
$\phi_S = 34°$

Note that ϕ computed from measurements taken along the meridian line is close to the latitude of Delos.

Deonna, p. 191, fig. 226-27 (drawing), pl. 542 (photograph).

3031G. Conical
Delos, Greece
Museum, storage room, no. 261

Worn rectangular gnomon hole is visible in the top surface of dial.

White marble - A trace of red color is visible on the engraved summer solstice and equinox of this dial. The worn winter solstice may also have been so emphasized. Eleven hour lines stretch between the solstices. Curling tendrils and a wide lip decorate the oblique front surface. On the base an egg and dart pattern flanks sprouting acanthus leaves and more tendrils.

Found on Delos.

H = 293 mm
W = 396 mm
D_{top} = 214 mm
D_{bottom} = 192 mm
tw = 58 mm
we = 52 mm
es = 100 mm
sb = 5 mm
(VI to VII)
w = 18.5 mm
e = 28.5 mm
s = 50 mm
$\sigma = 61°$
$\alpha = 60°30'$

For $\varepsilon = 24°$, $\omega = 35°30'$
g = 102 mm
$\phi = 26°12'$
$\phi_w = 27°$
$\phi_s = 40°42'$

Neither computed ϕ nor measured α is correct for Delos.

Deonna, p. 190, fig. 214 (drawing), pl. 532 (photograph).

3032. Conical
Delos, Greece
Unlocated in 1971

Lead remains in the gnomon hole at the top of the dial.

White marble - The hour lines and day curves are almost effaced. Upper corners are broken away. Two schematic feet flank the base.

$H = 250$ mm
$D_{top} = 115$ mm
$D_{base} = 100$ mm

Deonna, p. 190, plate 539 (photograph).

***

3033. Conical
Delos, Greece
Unlocated in 1971

The gnomon hole in the top surface is 13 mm deep.

White marble - Hour lines and day curves are effaced. Two schematic feet flank the base.

Found south of the temple of Artemis on Delos.

$H = 235$ mm
$W = 260$ mm
$D_{top} = 120$ mm
$D_{base} = 110$ mm
$90° - \alpha = 37°$
$\gamma = 54°$
$\sigma = 73°$

Deonna, pp. 190-91, figures 223-24 (drawings with measurements) and plate 544 (photograph).

3034G. Conical
Rhodes, Greece
Archeological Museum, study collection

Evidence of a rectangular gnomon hole (19 mm x 16 mm), containing a trace of lead, is preserved in the top of the dial.

Porous gray stone - Rounded sides rising from a plinth base characterize the general shape of this dial. There is a mount hole in the bottom surface. Ten preserved hour lines (the first line is broken away) are crossed by three parallel day curves. Only the meridian line continues lightly beyond the winter solstice. As the dial surface is damaged near the bottom, it is difficult to obtain an accurate measure of the segments of the summer solstice between hour lines and the segment of the meridian between the equinox and the summer solstice.

H = 327 mm
W = 460 mm
D_{base} = 210 mm

tw = 70 mm
we = 81 mm
es = 150+ mm

(VI to VII)
w = 29.5 mm
e = 49 mm
s = 84 mm

$\sigma = 72°$
$\gamma = 55°$

For $\varepsilon = 24°$, $\omega = 33°45'$, g = 162 mm,
and $\phi = 39°24'$, with $\phi_W = 37°$
and $\phi_S = 35°$.

Similar dials have been found on Rhodes (3035G, 3036G, 3037G) and Kos (3038G, 3039G, 3040G).

Unpublished

3035G. Conical Plate 35
Rhodes, Greece
Archeological Museum, study collection

Damage at the top surface of the dial has destroyed the dimensions of the gnomon hole.

White marble - Rounded sides rising from a plinth base mark this dial as a member of a group of similarly shaped examples from the islands of Rhodes and Kos. However, the usual conical shadow receiving surface characteristic of this group has, in this case, been slightly dished near the meridian. Parts of three parallel day curves and eleven hour lines are broken away on the left side. Only the meridian continues beyond the winter solstice to the gnomon, but all complete hour lines continue lightly beyond the summer solstice. Two parallel grooves decorate the lip. A trace of lead has been preserved in the square mount hole at the center of the bottom surface.

Found May 26, 1966, during construction on Mela Street, Rhodes.

For $\varepsilon = 24°$, $\omega = 29°$
$g = 138$ mm
$\phi = 37°48'$
$\phi_W = 45°12'$
$\phi_S = 46°42'$

See 1060G for a second dished conical surface.

Other dials with rounded sides and plinth base include 3036G, 3034G, 3037G, 3038G, 3039G, 3040G.

Unpublished

H = 220 mm
$\frac{1}{2}$W = 266 mm
D_{base} = 102 mm
tw = 55 mm
we = 70 mm
es = 117 mm
sb = 5 mm

(VI to VII)
w = 21.5 mm
e = 38.2 mm
s = 67 mm

$\sigma = 72.5°$
$\gamma = 52°$

chord = 186.5 mm (ws)
sagitta = 7 mm (at equinox)

34

35

3036G. Conical
Rhodes, Greece
Archeological Museum, garden

$H = 336$ mm
$W = 480$ mm
$D_{base} = 215$ mm
$tw = 65$ mm
$we = 63$ mm
$es = 149$ mm
$sb = 14$ mm
(VI to VII)
$w = 24$ mm
$e = 42.2$ mm
$s = 83$ mm
$\sigma = 79.5°$
$\gamma = 42.5°$

A rectangular gnomon hole 42 mm long, 33 mm wide, and 18 mm deep, containing no metal residue, is preserved in the top of the dial.

Porous grey stone - The general shape of the dial is characterized by rounded sides rising from a plinth base containing a round mount hole (diameter: 45 mm) in the bottom surface. Eleven hour lines are crossed by three parallel day curves. None of the hour lines continue beyond the winter solstice, but the meridian continues beyond the summer solstice and is traced onto the oblique front surface of the dial.

For $\varepsilon = 24°$, $\omega = 42°20'$
$g = 121$ mm
$\phi = 39°$
$\phi_W = 35°$
$\phi_S = 31°$

Dials with a similar shape have been found on Kos: 3038G, 3039G, 3040G.

Unpublished

3037G. Conical
Kamiros, Rhodes, Greece
Archeological Site on Rhodes

H = 454 mm
W = 2 x 370 mm
D = 295 mm

tw = 104 mm
we = 106.5 mm
es = 370 mm
sb = 15 mm

(VI to VII)
w = 40 mm
e = 68 mm
s = 140 mm

σ = 79°
γ = 47.5°

Traces of a gnomon hole 50 mm x 55 mm remain in the top surface to a depth of 40 mm.

White marble - Its monumental size distinguishes this dial from others of similar shape (rounded sides rising from a plinth base) found on the islands of Rhodes and Kos. The mount hole in the bottom center (45 mm x 50 mm, 55 mm deep) contains a trace of lead. The dial surface is broken away to the left of the meridian so that only six hour lines are preserved, none continuing beyond winter or summer solstice. The usual three day curves are parallel to each other and the lower edge. Four small holes have been drilled into the shadow-receiving surface: one at the juncture of meridian and winter solstice, one at the juncture of meridian and equinox, one along the equinox 96 mm from the meridian, and one along the summer solstice 163 mm from the meridian.

Found near a base in a public square at Kamiros.

For $\varepsilon = 24°$, $\omega = 51°12'$
$g = 163$ mm
$\phi = 36°$
$\phi_W = 17°$
$\phi_S = 16°$

Similar dials found on Rhodes include 3035G, 3036G, 3034G. Similar dials found on Kos include 3038G, 3039G, 3040G. See also 3058G from Ephesus.

Dyggve, E. and Poulsen, V., *Lindos, Fouilles de l'Acropole, 1902-1914 et 1952*, vol. 3, pt. 2, Copenhagen, 1960, p. 344, fig. 7, no. 24 (photograph of base).

3038G. Conical
Kos, Greece
Apotheke of the Archeological Museum

Traces of a deep gnomon hole and groove remain in top surface.

Marble - In this variation on the Rhodes-Kos design, the plinth base stretches necklike to rounded sides. All three parallel day curves are crossed by the eleven hour lines, but only the meridian continues all the way to gnomon and beyond the summer solstice onto the lip. Remains of a mount hole are preserved in the bottom surface.

H = 302 mm
W = 330 mm
D_{base} = 220 mm

tw = 44 mm
we = 44.5 mm
es = 86 mm
sb = 9 mm

(VI to VII)
w = 16 mm
e = 26.5 mm
s = 50 mm

σ = 83.5°
γ = 37.5°

For $\varepsilon = 24°$, $\omega = 35°30'$
$g = 87.6$ mm
$\phi = 34°42'$
$\phi_W = 35°30'$
$\phi_S = 44°18'$

Similarly shaped dials include 3034G, 3035G, 3036G, 3037G, 3039G, 3040G.

Unpublished

3039G. Conical
Kos, Greece
Apotheke of the Archeological Museum

Plate 36 left
Page 264

Wedge shaped gnomon hole-groove is 32 mm wide at back and 55 mm long.

Marble - The rounded sides of this dial rise from an elongated plinth neck. There is no mount hole in the bottom surface. All eleven hour lines continue beyond the winter solstice; only the meridian continues from the gnomon groove through the lip. Three day curves are parallel to the lower edge.

H = 308 mm
W = 290 mm
D_{base} = 132 mm

tw = 38 mm
we = 41 mm
es = 77 mm
sb = 4 mm

(VI to VII)
w = 14.2 mm
e = 24.2 mm
s = 47.5 mm

$\sigma = 81°$
$\gamma = 40°$

For $\varepsilon = 24°$, $\omega = 34°30'$
$g = 80.7$ mm
$\phi = 36°12'$
$\phi_W = 38°45'$
$\phi_S = 50°30'$

Dials with similar shape include 3034G, 3035G, 3036G, 3037G, 3038G, 3040G.

Unpublished

3040G. Conical
Kos, Greece
Archeological Museum, study collection in the Castle of the Knights of St. John

Plate 36 right
Page 264

The gnomon hole in the top of this dial is rectangular (38 mm x 18 mm), 33.5 mm deep, and free of any trace of metal.

Marble - The general shape of the dial is distinguished from that of the majority of ancient sundials by rounded sides which rise out of a square plinth-like base. There is a square mount hole in the bottom of the base. On the shadow-receiving conical surface eleven hour lines cross three day curves, extending beyond both the winter and the summer solstice; only the meridian continues all the way to the gnomon hole.

Found during Italian excavations of Kos.

H = 280 mm
W = 390 mm
base = 175 mm x 185 mm

tw = 56.5 mm
we = 57.5 mm
es = 101.5 mm
sb = 15 mm

(VI to VII)
w = 22.2 mm
e = 35 mm
s = 58.8 mm

$\sigma = 67.8°$
$\gamma = 59°$

For $\varepsilon = 24°$, $\omega = 32°$
$g = 121$ mm
$\phi = 37°10'$
$\phi_W = 33°30'$
$\phi_S = 36°42'$

Six similar dials have been found on Kos (3039G, 3038G) and Rhodes (3035G, 3036G, 3037G, 3034G).

Unpublished

3041. Conical
Chios, Greece
Chios Museum

A gnomon groove is visible above the meridian line.

H = 440 mm
W = 280 mm

White marble - Eleven hour lines radiating from the gnomon groove divide three engraved day curves. Two lion heads rest on lion feet at the left and right of the acanthus leaf on the base. A relief lip follows the front of the dial surface.

Found in the ruins of a Byzantine basilica on Chios.

Dial 1056G, now in the British Museum, has a similarly ornamented base.

Archaiologikon Deltion, Chronika 21(1969): 385 and plate 406ζ (photograph).

3042. Conical
Melos, Greece
Archeological Museum

White marble - Five hour lines (2,3,4,5,6,) and three day curves are visible in a photograph taken by the Ephor of Myconos. A relief margin borders both sides. A disk in relief is also visible in the side view. Two stylized legs flank the base in front.

we = 40 mm
es = 66 mm

(VI to VII)
w = 17 mm
e = 27 mm
s = 42 mm

Description from photograph in file of Ephoreia of Myconos.
Measurements taken by Gerry Gesell in May, 1971.

Unpublished

***

3043. Conical
Melos, Greece
Archeological Museum

White marble - Parts of five hour lines (6,7,8,9,10) are visible in a photograph taken by the Ephor of Myconos; parts of three day curves can also be seen. The base is simple and unornamented.

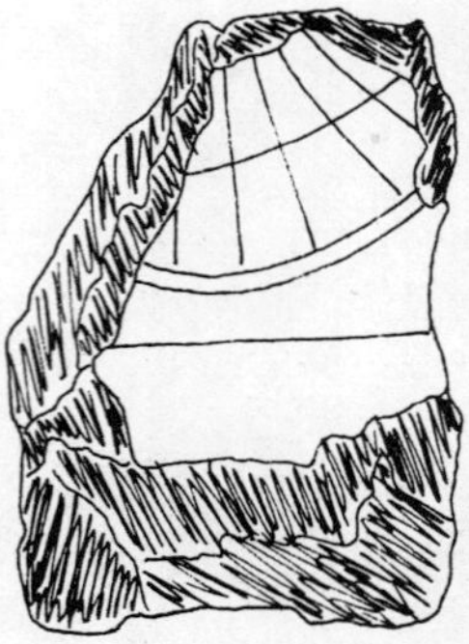

Unpublished

3044G. Conical
Athens, Greece
Epigraphical Museum, no. 2922

Plate 37
Page 264

A round gnomon hole in the top surface contains ferrous stains.

H = 270 mm
W = 300 mm
D = 170 mm

σ = 66°
γ = 62.5°
β = 38+°

White marble streaked with gray - Eleven hour lines of varying lengths radiate from an area near the gnomon to an area near the lower edge. Two lines of Greek inscription have been engraved on the lowest part of the cyma base:

Λυσαγόρας Πρασίνου(?) Φυλάσ[ιος]
Θεοῖς Μεγάλοις

Possible provenance: Imbros.

Dated by epigraphy (Α Θ Μ Σ) to the second or first century B.C.

The Θεοὶ Μεγάλοι are the Dioskouroi-Kabeiroi, earth gods of ancient Imbros.

For other conical dials with hour lines only see 3003G, 3017G, 3018G, 3066G, 3088, 3091, 3099.

Foucart, P., "Inscriptions des Clerouques Athéniens d'Imbros," BCH 7(1883): 164.
Graindor, P., "Inscriptiones Grecques," RA 6(1917): 1 and figure 1 (photograph).
IG, XII, 8, 71.

36

37

3045. Conical
Kephalos, Greece

Marble - Only seven hour lines are preserved on the much damaged shadow-receiving surface. The only visible day curve seems to be the equinox. In the relief design on the base, two male figures stand facing each other holding flowers above a vase.

The dial fragment was found during excavation of a early Christian basilica on Kephalos, a small island in Amvrakikos Bay on the western coast of Greece.

To Ergon, 1966, p. 91f, figure 108 (photograph).

***

3046. Conical
Samos, Greece
Tignani-Kastro Magazine, no. 323 (lost)

H = 330 mm
W = 365 mm
D = 220 mm

White marble - Eleven hour lines extend from the upper edge across the engraved winter solstice and equinox to the summer solstice. The dial is apparently marked with a daylight triangle which Tölle has drawn incorrectly. The meridian is marked on the center front of the base.

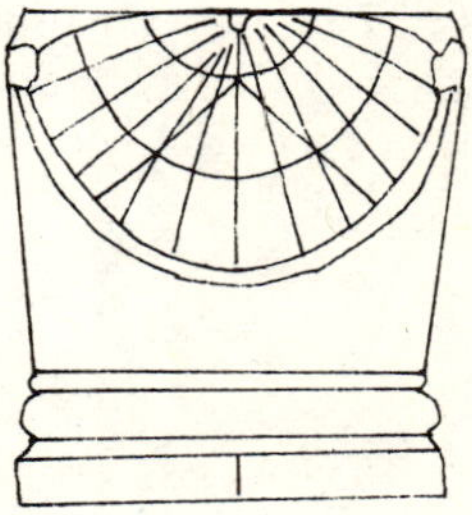

Found in the vineyard of Stephanos Kodsakos about 1910.

Tölle, Renate, "Uhren auf Samos," in *Opus Nobile*, Wiesbaden, 1969, pp. 164-71, figure 10 (drawing).

3047. Conical
Samos, Greece
Archeological Museum, no. 322

H = 330 mm
W = 385 mm
D = 260 mm

The gnomon hole in the top surface of the dial is visible above the meridian.

White marble - Eleven hour lines stretch across the equinox between engraved solstice surves. Day curves are identified by Greek inscription:

χειμερινὴ τροπή
[ἰ]σημερινὴ τροπή
θερινὴ τροπή

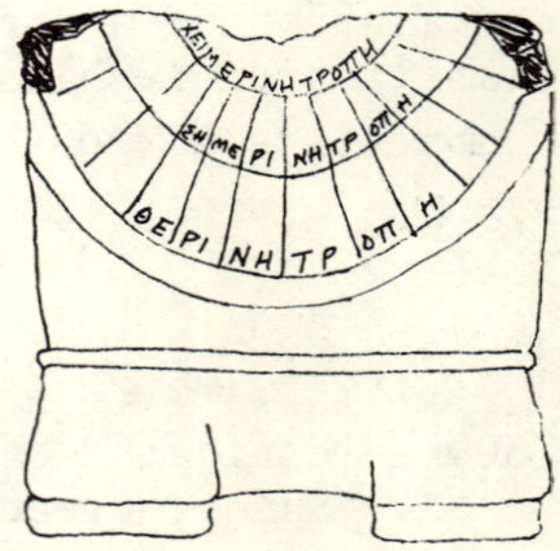

Dated by epigraphy to the second half of the second century B.C. (P. Herrmann in Tölle).

See 3060G for a similar design.

Theophanides, B. D., *Archaiologikon Deltion* 12 (1929): 236-37 (photograph).
Tölle, R., "Uhren auf Samos," in *Opus Nobile*, Wiesbaden, 1969, pp. 164-71, figure 11 (drawing).

3048. Conical
Paphos, Cypros
Paphos Museum, no. MTT1504

Plate 38
Page 273

A much damaged bronze gnomon is in place at the top of the dial above the meridian.

Limestone - Four day curves have been engraved on the conical surface always equidistant from each other and from the lower edge. Eleven hour lines evenly divide the day curves. A simple cyma molding decorates the base.

tc_1 = 65 mm

c_1c_2 = 76 mm

c_2c_3 = 83 mm

c_3c_4 = 62 mm

(VI to VII)
c_2 = 41 mm

$\alpha = 60°$

For $\varepsilon = 24°$, tw = 65 mm, we = 76 mm, and es = 145 mm: $\omega = 49°36'$, g = 104 mm, and $\phi = 28°18'$; c_2 corresponds to $\delta = -11°36'$.

Measurements taken by Derek J. de Solla Price.
Photograph from the Department of Antiquities, Nicosia, Cyprus.

3049G. Conical Combination
Paris, France
Louvre Museum, no. 2820

H = 302 mm
W = 445 mm
D = 405 mm

White marble - North and south-facing conical surfaces have been carved into this marble rhombohedron with oblique angles of 128°. A Greek inscription below the south-facing surface reads:

Βασιλεῖ Πτολεμάιῳ Ἀπολλώνιος Ἀπολλοδότου
Θεμισταγόρας Μενίσκου Ἀλεξανδρευς ἐποίει

Found in 1873 at Heraclea ad Latmum in Turkey.

A similarly designed dial is preserved in the National Archeological Museum, Palermo, Italy (see 7003G).

South Face

Of eleven hour lines, only the meridian continues beyond the summer solstice to the lower edge of the conical surface. These hour lines evenly divide seven day curves in all. Workmanship is very fine.

For $\varepsilon = 24°$, $\omega = 32°$, $g = 123$ mm, and $\phi = 35°$.

For $\omega = 25°$, $\phi = 38°$, with $\phi_W = 30°$ and $\phi_S = 26°$.

$tc_1 = 66.5$ mm
$c_1c_2 = 14$ mm
$c_2c_3 = 10$ mm
$c_3c_4 = 17.5$ mm
$c_4c_5 = 32$ mm
$c_5c_6 = 40$ mm
$c_6c_7 = 19$ mm
$c_7b = 55$ mm
(VI to VII)
$c_1 = 23$ mm
$c_2 = 26.5$ mm
$c_3 = 29$ mm
$c_4 = 35$ mm
$c_5 = 42$ mm
$c_6 = 50$ mm
$c_7 = 55.2$ mm
$\sigma = 63°$
$\gamma = 65°$
$\alpha = 52°$

3049G. continued

North Face

Of eleven hour lines, only the meridian continues above the winter solstice to the gnomon hole and beyond the lower edge of the conical surface to the lower edge of the dial. Two engraved day curves are visible.

ts = 68 mm
se = 24 mm
eq = 18 mm
qb = 128 mm

(VI to VII)
s = 17.2 mm
e = 23.5 mm
w = 59.0 mm

Rayet concludes that the shadow-receiving surface of this north face is a truncated cone with opening about 80° and height 21 mm; the base circles have diameters of 420 mm and 168 mm. The hour lines are effective only between equinox and summer solstice.

Rayet, pp. 61-66 and plate 1 (drawings).
Cf. Diels, pp. 176-77, figs. 55, 56, 57 (drawings).
Gatty, pp. 37-38 (drawing).
Loewy, E., *Inschriften Griechischer Bildhauer*, Leipzig, 1885, p. 298.
DA, vol.3, p. 259, figure 3886 (drawing).

***

3050G. Conical
Aphrodisias, Turkey
Excavation depot, no. 1964 64-415

Coarse-grained, grayish Aphrodisias marble - This fragment is the upper left corner of a conical dial, apparently designed with rounded back. Four hour lines and inscribed equinox are preserved.

ἰ̓[σ]η[μερινή]
θερ[ινή]

H = 145 mm
W = 230 mm
D = 40 mm
(I to II)
s = 49.5 mm
e = 40 mm
γ = 60°
α = 52°

Unpublished

3051G. Conical
Claros, Turkey
In situ north of the altar of the Temple of Apollo

A rectangular gnomon hole, 52 mm long x 45 mm wide x 78 mm deep, is preserved in the top surface of this dial. A 28.5 mm long groove connects the hole with the top edge.

White marble - Eleven fine hour lines and three day curves form the dial grid on this worn conical surface. Above the winter solstice, the lines continue very lightly to the gnomon groove. The oblique front ends in a cyma base. Workmanship is especially good.

H = 788.5 mm
W = 110 mm
D_{top} = 752 mm
D_{bottom} = 500 mm
tw = 155 mm
we = 148 mm
es = 249 mm
sb = 5 mm
(VI to VII)
w = 58.6 mm
e = 93 mm
s = 158.2 mm
$\sigma = 71.2°$
$\gamma = 60°$
$\alpha = 50°$

For $\varepsilon = 24°$, $\omega = 30°$
g = 332 mm
$\phi = 37°48'$
$\phi_W = 36°$
$\phi_S = 42°36'$

Akurgal reports the existence of an inscription on this dial stating that it was donated by the "agoranomos" in Hellenistic times. I could not find any trace of inscription, but the base is now under several feet of water and buried in mud.

Cf. Akurgal, Ekrem, *Ancient Civilizations and Ruins of Turkey*, Istanbul, 1970, p. 139.
Martin, Roland, *Manuel d'Architecture Grecque*, Paris, 1976 (photograph).

3052. Conical
Knidos, Turkey

A gnomon was apparently attached to the top surface of the dial in a series of three holes.

H = 710 mm
W = 570 mm
D = 530 mm

tw = 65 mm
we = 60 mm
es = 88 mm
sb = 22 mm

White marble - Three day curves on the conical surface are divided by eleven hour lines. The dial appears from photographs to be very well made. Stylized lion feet decorate the base. Part of a simple relief design including a curling leaf and small flower decorates the upper half of the right side.

This monumental dial was found at the foot of the east steps of a Roman Corinthian temple during excavations at Knidos in July and August, 1967.

For $\varepsilon = 24°$, $\omega = 23°$

Mellink, M. J., "Archaeology in Asia Minor." AJA 72 (1968): 138, plate 56.13 (photograph).

3053G. Conical Plate 39
Paris, France
Louvre Museum, no. MA 2880

A portion of bronze gnomon remains in position in the top of the dial above the meridian.

White marble - Eleven hour lines radiate from the gnomon to the lower edge of the conical surface. Two day curves, always equidistant from each other, cross the hour lines. These day curves are not always equidistant from the lower edge. Two four-toed griffon feet decorate the base. Honeysuckle vines and flowers decorate the center of the base and the left and right sides.

Found at Pergamon.

H = 551 mm
PH = 313 mm
W = 196 mm
D = 171 mm

tw = 35 mm+
we = 52.5 mm
eb = 9 mm

(VI to VII)
w = 11 mm
e = 24 mm
b = 28 mm

σ = 52.2°
γ = 73.2°
α = 53.2°
β = 38°

Dial 1004 from Loutra in Thrace has similar large griffon legs and feet.

de Clarac, F., *Musée de Sculpture*, Paris, 1841-50, plate 254, no. 800.
Cf. Ardaillon, p. 257, figure 3884 (drawing).

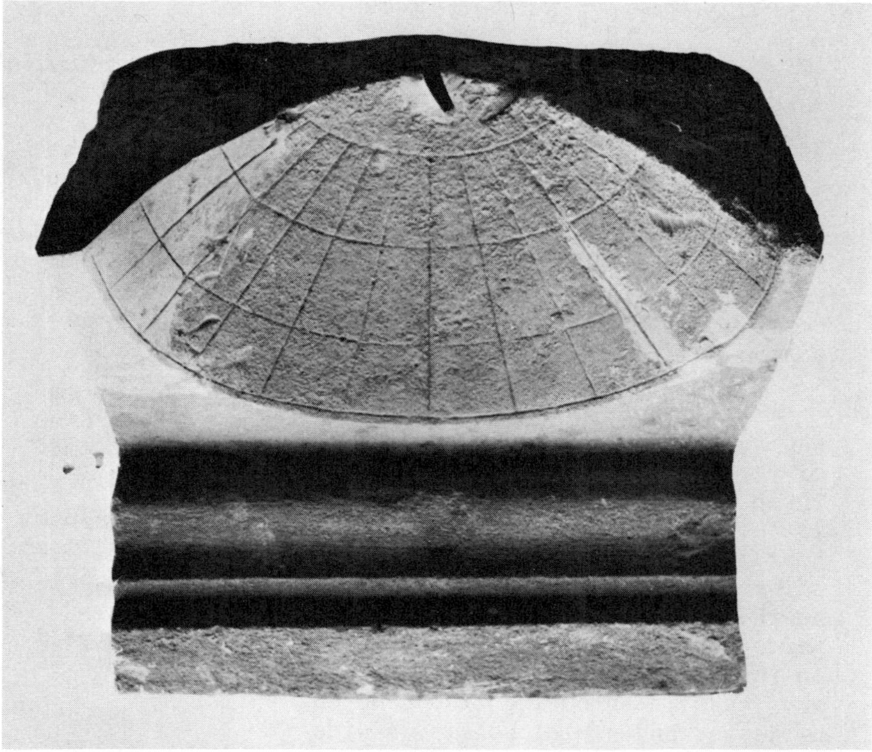

38

39

3054G. Conical
Bergama, Turkey
Archeological Museum, portico

Plate 40
Page 279

A wide rounded depression has been cut lengthwise in the top surface, destroying all but a trace of the gnomon hole.

White marble - The depression in the top surface has destroyed parts of day curves and hour lines, but three of the former and eleven of the latter are visible. The length of the hour lines is limited by the solstice curves. A figure in high relief in the pose of the Farnese Atlas supports the dial surface. A low relief on the sides at the top depicts stems and tendrils sprouting from a bunch of leaves and eyelet petals. A wide lip follows the front curve.

Found at ancient Pergamon.

H = 785 mm
W = 359 mm
D_{top} = 290 mm

tw = 29.5 mm
we = 29 mm
es = 136 mm
sb = 6 mm

(VI to VII)
w = 13 mm
e = 21 mm
s = 57.1 mm

$\sigma = 69°$
$\gamma = 53.2°$
$\beta = 37°$

For $\varepsilon = 24°$, $\phi_W = 4°$
$\phi_S = 8°48'$

See catalogue entry 1034 for a drawing of a dial with similar supporting figure.

Unpublished

3055G. Conical
Berlin, East Germany
Pergamon Museum, Antiken Sammlung

White marble - The shadow-receiving surface is broken away above the equinox so that only the equinox and the summer solstice are preserved. Seven hour lines (3,4,5,6,7,8,9) are visible. None continue beyond the summer solstice. A lip 17 mm in diameter follows the curve of the circular section in front. Part of an inscription in Greek capital letters is preserved on the base:

['Ε]λπιδήφορος Ἀγαθίου Μοσχίων
[--------]των τὸ ὡροσκόπιον

PH = 437 mm
W = 580 mm
D_{max} = 550 mm
te = 153.5 mm
es = 142 mm
sb = 11 mm

(VI to VII)
e = 51.6 mm
s = 88.5 mm

β = 37°
γ = 55.5°

Provenance unknown.

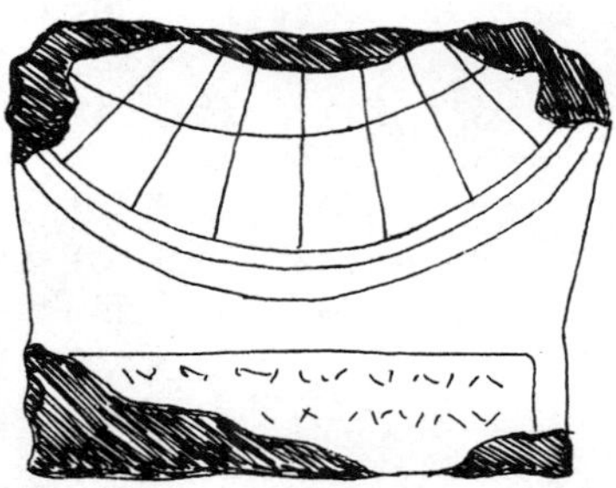

For ω = 34.5°, ϕ = 37°

Unpublished

3056G. Conical — Plate 41, Page 279
Selçuk, Turkey
Archeological Museum, depot, no. 374

$H = 238$ mm
$W = 315$ mm
$D_{base} = 258$ mm

$tw = 29.5$ mm
$we = 26$ mm
$es = 57$ mm
$sb = 2$ mm

(VI to VII)
$w = 14.3$ mm
$e = 20$ mm
$s = 33$ mm

$\sigma = 60°$
$\gamma = 67°$
$\beta = 37.5°$

A rectangular gnomon hole in the top surface (29 mm long x 15 mm wide x 23 mm deep) contains some traces of lead.

White marble with gray - Broken sides have destroyed hour lines one and eleven. The nine remaining lines extend from the winter solstice across the equinox and the summer solstice to the lower edge. A wide lip emphasizes the lower edge. Duck-pin shaped legs on the base end in widespread four-toed feet. Petals and curling tendrils sprout from a calyx on the side reliefs. A square mount hole is visible in the bottom center.

Found on the slope of the hill of the Basilica of St. John near Ephesus.

For $\varepsilon = 24°$, $\omega = 40°$
$g = 57$ mm
$\phi = 45°24'$
$\phi_w = 3°36'$
$\phi_s = 6°48'$

Unpublished

3057G. Conical
Selçuk, Turkey
Archeological Museum, depot, no. 1391

A depression 146 mm wide and 25.5 mm deep has been cut into the top surface of this dial. In the center of this depression, a rounded groove 60 mm long and 28 mm wide leads from the back of the dial to the front edge above the meridian.

Cream marble - The meridian extends from the gnomon groove to the front lower edge. The other ten hour lines extend from the winter solstice across the equinox to the lower edge. Four-toed lion paws flank the base. A cylindrical mount hole (30 mm in diameter and 54 mm deep) pierces the bottom surface, and a clamp hole and groove mark the left side.

Found during construction of the Tusan Motel near Ephesus, on August 3, 1961.

For $\varepsilon = 24°$, $\omega = 14°$, $\phi_W = 53°$
and $\phi_S = 56°42'$.

It seems likely that the lower edge does not mark the summer solstice since the measured length eb is less than the measured length we and this is not an antiboreal dial.

H = 292 mm
W = 344 mm
D_{base} = 206 mm
tw = 70 mm
we = 62.5 mm
eb = 50 mm

(VI to VII)
w = 22.3 mm
e = 33.3 mm
s = 44.2 mm

$\sigma = 58.5°$
$\gamma = 78°$
$\beta = 39°$

Unpublished

3058G. Conical — Plate 42
Vienna, Austria
Kunsthistorisches Museum, no. I1639

The horizontal section of the gnomon hole is a narrow rectangle 55 mm long. The depth of the hole is 94 mm.	H = 977 mm W = 1320 mm D = 140 mm
Marble - Three fragments have been joined. In all seven hour lines (3,4,5, 6,7,8,9) and three day curves are preserved. Hour lines do not extend beyond either solstice. Portions of a Greek inscription mark the winter solstice: [χε]ιμερ[ινή] equinox: [ἰσημ]ερῑν[ή] summer solstice: [θε]ρινή	tw = 194 mm we = 302 mm es = 372 mm sb = 27 mm (VI to VII) w = 86 mm e = 165 mm s = 258 mm
Found during excavation of Ephesus some time before 1907.	$\sigma = 70°$ $\gamma = 54°$

For $\varepsilon = 24°$, $\omega = 13°$, $g = 709$ mm
and $\phi = 47°$, with $\phi_W = 54°$
and $\phi_S = 53°$

This partially preserved dial is the largest known. In general shape, it resembles dials found on Rhodes and Kos (cf. 3034G).

Noll, R., *Griechische und Lateinische Inschriften der Wiener Antiken Sammlung*, Vienna, 1962, p. 36, no. 64.

40

41

42

3059G. Conical
Side, Turkey
Museum, no. 1391

The gnomon groove in the top surface is 101 mm long.

White marble - Eleven hour lines extend between an engraved day curve near the top of the dial and the lower edge. The design is very simple.

Found at Side.

H = 247 mm
W = 428 mm
D_{bottom} = 192 mm
PD_{top} = 230 mm
cb = 124 mm
(VI to VII)
c = 16 mm
b = 42 mm
$\sigma = 70°$
$\gamma = 57°$

The lower front edge of the dial face between the fourth and the ninth hour lines is divided into arcs measuring: 44, 44, 42, 42, 51.5 mm.

For $\omega = 33°$, tw = 46.9 mm
we = 44 mm
es = 80 mm
g = 95 mm
$\phi = 37°$

This dial is a unique example of a conical dial having only solstitial day curves. Spherical dials with only solstice curves include 1067G, 1042, 1041, 1011.

Unpublished

3060G. Conical
Izmir, Turkey
Archeological Museum, no. 5404

Plate 43
Page 288

The gnomon hole in the top surface (23 mm long and 15 mm wide) still contains a trace of lead.

Gray marble - Eleven finely engraved hour lines extend from the winter solstice across the equinox and the summer solstice to the lower edge. Slightly stylized three-toed feet with claws decorate base. The day curves are identified with a Greek inscription.

χειμερινός
ἰσημερινός
θερινός

Found at Aspira?

H = 310 mm
W = 360 mm
D = 190 mm

tw = 46.5 mm
we = 70 mm
es = 79.5 mm
sb = 11 mm
(VI to VII)
w = 18.2 mm
e = 35.5 mm
s = 55.5 mm

$\sigma = 68°$
$\gamma = 57°$
$\beta = 38°$

For $\varepsilon = 24°$, $\omega = 8°$
$g = 162$ mm
$\phi = 47°$
$\phi_W = 56°$
$\phi_S = 56°$

Conical dial 3058G from Ephesus has similar small cone opening and discrepancy between measured β and computed ϕ. In both 3058G and 3060G, we is very nearly equal to es.

Unpublished

3061G. Conical
Ostia Antica, Italy
Magazine, no. 15903

The gnomon hole is 15 mm deep.

Travertine marble - The surface of the dial is quite worn and the right side is broken away, but seven hour lines (1,2,3,4,5,6,7) affect an irregular division of the shadow-receiving surface. Only the summer solstice is clearly visible but the positions of other day curves can be estimated from preserved traces. A half-cylinder molding separates the oblique front from the sloping base.

Found while cleaning a small temple in the area of the four temples at Ostia Antica.

H = 350 mm
W = 336 mm
D_{top} = 242 mm
D_{base} = 123 mm
tw = 46.5 mm
we = 47.5 mm
es = 68.5 mm
sb = 8.5 mm
(V to VI)
w = 19.5 mm
e = 34 mm
s = 53 mm
$\sigma = 71^\circ$
$\gamma = 60^\circ$
$\alpha = 49^\circ$
$\beta = 41.2^\circ$

For $\varepsilon = 24^\circ$, $\omega = 22^\circ$
g = 129 mm
$\phi = 47^\circ 24'$
$\phi_w = 47^\circ 30'$
$\phi_s = 43^\circ 24'$

Other dials with cylindrical molding include 3072G, 1028G, 1032G, 1021G.

Unpublished

3062G. Conical
Ostia Antica, Italy
Magazine, no. 1498

A trace of lead fill remains in the gnomon hole.

H = 253 mm
W = 285 mm
D = 127 mm

tb = 135 mm

σ = 74°
γ = 63°
β = 43.2°

Large-grained Greek marble - No engraved lines are preserved on the shadow-receiving surface. An 18 mm lip follows the lower edge of this conical surface. Stylized legs flank the base. A relief strip 30 mm wide crosses the left and right sides diagonally, meeting the back at an angle of 43.2°.

Found on the SE side of the junction of the Decumanus and the plaza of the Altar of the Lares at Ostia Antica.

Unpublished

3063G. Conical
Ostia Antica, Italy
Magazine, no. 14382

Part of a bronze gnomon and lead adhesive remain in the top of the dial.

Stained white marble - A round mount hole in the bottom surface has worn through to the top of the base. None of eleven hour lines continue beyond the solstices. Three day curves are always equidistant from each other and the lower edge of the shadow-receiving surface.

H = 300 mm
W = 292 mm
D_{top} = 256 mm
D_{base} = 92 mm
tw = 33 mm
we = 43.5 mm
es = 86 mm
sb = 3 mm
(VI to VII)
w = 17 mm
e = 27 mm
s = 47 mm
$\sigma = 77°$
$\gamma = 52.2°$
$\alpha = 48°$
$\beta = 40°$

For $\varepsilon = 24°$, $\omega = 36°30'$
g = 84 mm
$\phi = 43°$
$\phi_W = 30°$
$\phi_S = 31°$

Unpublished

3064G. Conical
Pompeii, Italy
House of the Golden Cupids,
VI,16,7, peristyle

A complete gronze gnomon remains in a lead-filled hole in the top surface of the dial.

White marble - Lion legs and feet flank the base of this dial. A relief margin outlines the edges of the dial at the sides. Within the margin is a relief consisting of a calyx, from which sprouts a stem ending in a second calyx. Eleven hour lines cross three day curves but do not extend beyond either solstice.

The dial has been placed in the garden of the House of the Golden Cupids. It is probably not in its original position as it does not face south.

The House of the Golden Cupids dates from the Neronian period.

For $\varepsilon = 24°$, $\omega = 34°30'$
$g = 88$ mm
$\phi = 52°$
$\phi_w = 29°12'$
$\phi_s = 29°12'$

For tw = 38 mm, $\phi = 45°36'$.

Other conical dials with preserved gnomon include 3001G, 3024G, 3048, 3023G. Similar side reliefs decorate 3026G, 3071G, and 3079G.

H = 310 mm
W = 330 mm
D_{top} = 226 mm

gw = 27 mm
we = 38 mm
es = 71.5 mm
sb = 7 mm

(VI to VII)
w = 18 mm
e = 28 mm
s = 46.8 mm

$\gamma = 59°$
$\beta = 38°$

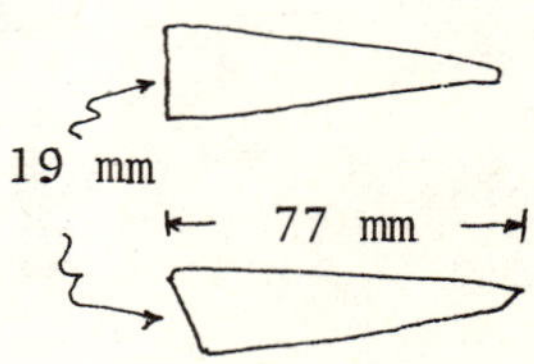

gnomon

Mauri, Amedeo, *Pompeii*, Rome, 1970, p. 147, plate 33 (photograph of dial in position).

3065G. Conical
Pompeii, Italy
Granario

A 17 mm deep gnomon hole is preserved in the top surface of the dial.

Gray porous stone - Traces of eleven hour lines are limited by the winter solstice and the lower edge of the shadow-receiving surface. An engraved equinox line is also preserved. Two narrow legs with cyma curved profiles flank the base. The round mount hole in the bottom surface has worn through to the top of the base at right of center.

For $\varepsilon = 24°$, $\omega = 10°$, $g = 110$ mm, and $\phi = 38°48'$, with $\phi_W = 50°36'$ and $\phi_S = 56°30'$.

H = 306 mm
W = 292 mm
D_{top} = 165 mm
D_{base} = 153 mm
tw = 32 mm
we = 55 mm
es = 64 mm
sb = 7.5 mm

(VI to VII)
w = 13 mm
e = 22 mm
s = 35 mm

$\alpha = 68°$
$\sigma = 63°$
$\beta = 22.5°$
$\gamma = 54°$

Unpublished

3066G. Conical
Pompeii, Italy
Granario

Plate 44
Page 288

A 103 mm long bronze gnomon (16 mm wide at the base and tapering to a point) is in position at the top of the dial. No lead fill is visible.

White marble streaked with gray - No day curves cross the eleven deeply engraved hour lines. Two lines, always 35 mm distant from each other, engraved in the oblique front surface of the dial emphasize the lower edge of the shadow-receiving surface. Two legs with curving profiles flank the base.

H = 383 mm
W = 550 mm
D_{top} = 308 mm

gb = 221 mm

(VI to VII)
b = 76 mm

σ = 80°
α = 50°
γ = 29°

For other conical dials with hour lines only see: 3033G, 3017G, 3018G, 3044G, 3088, 3091, 3099, 3103.

Unpublished

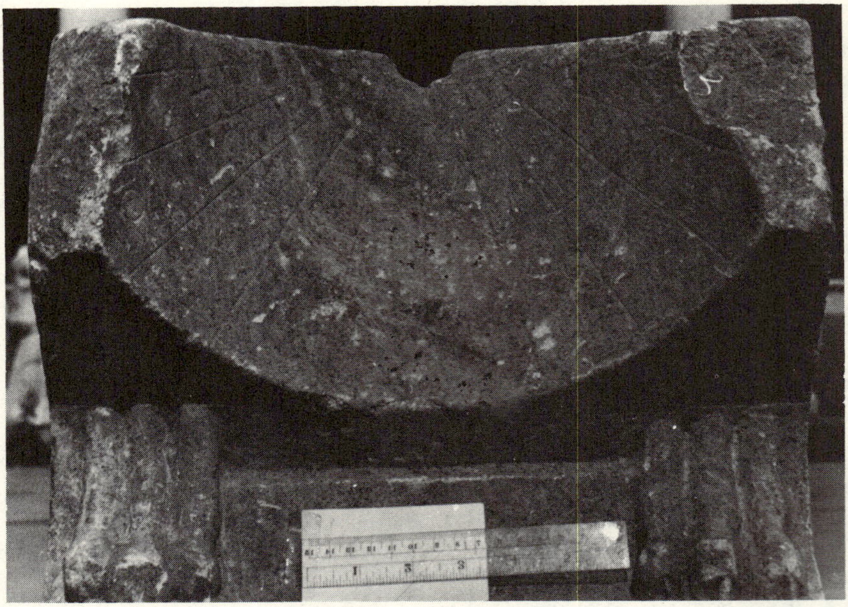

43

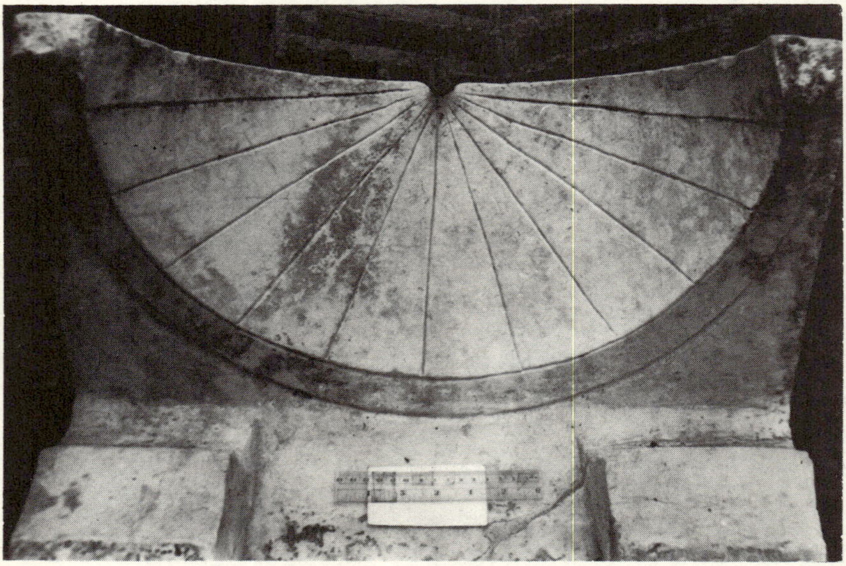

44

3067G. Conical — Pompeii, Italy — Granario — Plate 45, Page 292

Lead fills the gnomon hole in the top surface of the dial. An 8 mm deep groove with sloping sides connects the hole with the front surface.

White limestone - The eleven hour lines do not continue beyond the winter solstice but continue lightly beyond the summer solstice. These two solstices are always equidistant from each other and from the lower edge. The third, sixth, and ninth hour lines are marked as shown in the plate. A crescent flanked by two six-leafed rosettes decorates the oblique front of the dial below the shadow-receiving surface. A half-cylinder molding separates this decorated face from the sloping base. Construction guide lines remain on the right side of the dial as shown in the diagram.

H = 255 mm
W = 241 mm
D_{top} = 172 mm
tw = 30 mm
we = 36.5 mm
es = 51 mm
sb = 5 mm

(VI to VII)
w = 12 mm
e = 21 mm
s = 35 mm

$\alpha = 49°$
$\sigma = 80°$
$\gamma = 51°$

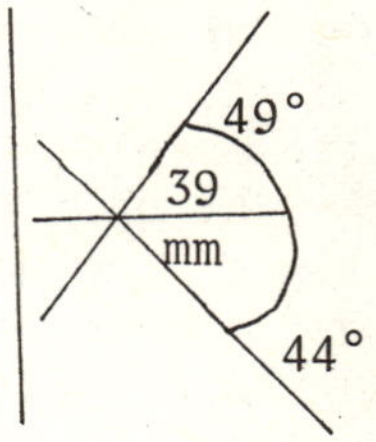

For $\varepsilon = 24°$, $\omega = 20°30'$, g = 84 mm, and $\phi = 42°18'$, with $\phi_W = 48°18'$ and $\phi_S = 52°18'$.

The third, sixth, and ninth hour lines are also specially marked on 1016G, 1022G, 1056G, 2023G, 3022, 3076, 3080G.

A rose and crescent design have also been engraved on the front surface of 3022, 1037G, and 1059.

Unpublished

3068. Conical
Pompeii, Italy
Lost in storage area

White marble - The left half of the dial, including any trace of gnomon hole is broken away. Parts of seven hour lines (5,6,7,8,9,10,11) are preserved. Two engraved day curves are visible parallel to the lower edge. Two lion legs and feet flank a base that shows traces of an obliterated inscription.

tw = 19 mm
we = 28 mm
es = 55 mm

$\alpha = 42°$
$\sigma = 90°$

For $\varepsilon = 24°$, $\omega = 36°10'$.

Measured by D. J. de Solla Price.

***

3069. Conical
Pompeii, Italy
Lost in storage area

Part of gnomon hole is preserved in the top surface of the dial.

White marble - This fragment is the upper right half of a conical dial. Six hour lines, including the meridian (6,7, 8,9,10,11), and parts of three engraved day curves are preserved.

H = 180 mm

we = 40 mm
es = 113 mm

$\alpha = 50°$
$\gamma = 60°$

For $\varepsilon = 24°$, $\omega = 47°$.

Measured by D. J. de Solla Price.

***

3070. Conical
Pompeii, Italy
Unlocated in 1971

A photograph captioned "Sundials at Pompeii" published in Gabba, L., and Jacono, L., "Meridiana," *Enciclopedia Italiana*, p. 903, shows a complete conical sundial which cannot be identified with other catalogue entries.

3071G. Conical
Naples, Italy
National Archeological Museum, no. 2541

Plate 46
Page 292

A lead-filled gnomon hole is visible in the top surface of the dial.

Travertine marble - Of the eleven hour lines, only the meridian extends from the gnomon groove to the lower edge of the conical surface; although some other hour lines do seem to extend beyond the summer solstice. In addition to the summer solstice, both the equinox and the winter solstice are represented by engraved lines. Sprouting calyx, flower and curling tendril are carved in low relief on the sides of the dial. Two lion feet flank a base inscribed with the following Oscan inscription:

Mr. Atiniis Mr Kvaisstur eitiuvad
multasikad Kombennieis tangi[n(ụd)]
amanaffed

Found near the Stabian baths in Pompeii in 1854.

The Stabian baths date from the second century B.C.

For $\varepsilon = 24°$, $\omega = 42°$
$g = 103$ mm
$\phi = 42°$
$\phi_W = 27°$
$\phi_S = 41°$

Similar side reliefs decorate 3026G, 3079G, and 3064G.

H = 430 mm
W = 505 mm
D_{base} = 230 mm
tw = 48.5 mm
we = 56 mm
es = 132 mm
sb = 8 mm

(VI to VII)
w = 22.2 mm
e = 36.5 mm
s = 76.5 mm

$\sigma = 85°$
$\beta = 33.5°$

Quaranta, B., *L'Orologio a Sole di Beroso*, Naples, 1854.
Zvetaiff, Ivan, *Inscriptiones Italiae Inferioris Dialecticae*, Moscow, 1886, p. 52, no. 146.
Cf. Diels, pp. 175, 176, and plate 14, 2 (drawing).

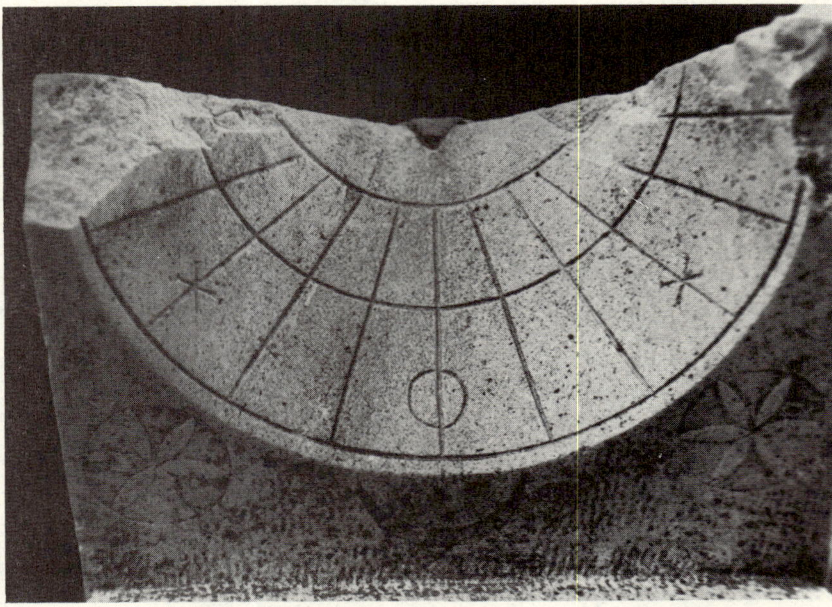

45

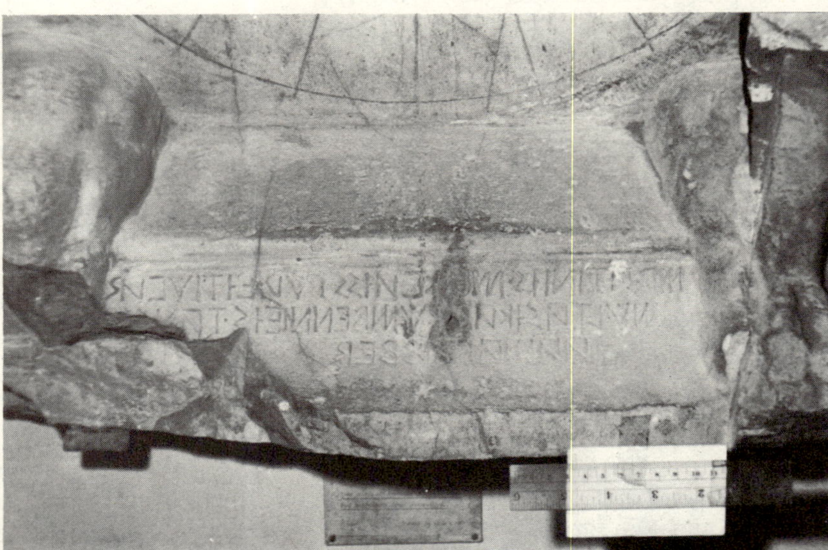

46

3072G. Conical — Plate 47
Naples, Italy — Page 305
National Archeological Museum, garden

Lead fill remains in the gnomon hole at the top of the dial.

White marble - Three day curves are divided unevenly by eleven hour lines. A simple engraved design decorates the oblique front surface. Below the design, a rolled scroll molding fills the angle between the simple base and the oblique front.

Found July 14, 1832, during excavation of the peristyle of the House of the Figured Capitals in Pompeii.

$H = 432$ mm
$W = 356$ mm
$D_{top} = 350$ mm
$D_{base} = 220$ mm
tw = 51 mm
we = 58 mm
es = 72 mm
sb = 3 mm
(VI to VII)
w = 20 mm
e = 36.2 mm
s = 55 mm
$\sigma = 86.5°$
$\alpha = 46°$
$\gamma = 41.5°$

The House of the Figured Capitals dates from the period between 200 and 80 B.C.

For $\varepsilon = 24°$, $\omega = 13°15'$
$g = 155$ mm
$\phi = 46°45'$
$\phi_w = 52°6'$
$\phi_s = 50°$

Dials with similar geometric design include 3085G and 3076.

Avellino, F. M., *Descrizione di una Pompeiana*, Naples, 1837, plate 10, 12 (drawing) and pp. 32 and 60.
Cf. Ardaillon, p. 259, figure 3885 (drawing).
Bilfinger, G., *Die Zeitmesser der Antiken Völker*, Stuttgart, 1886, p. 27, figure 2 (drawing).
Deonna, BCH 56(1932): 470.

3073G. Conical
Naples, Italy
National Archeological Museum, garden

The base of a bronze gnomon remains in lead fill at the top of the dial.

White marble - Eleven hour lines extend from the gnomon groove to the lower edge of the conical surface. Three day curves cross these hour lines. A 52 mm wide lip emphasizes the circular section in the oblique front of the dial. Two four-toed lion feet flank the base. The left and right sides meet at the back of the dial forming a dihedral angle equal to 88°.

H = 540 mm
W = 440 mm
$D_{top\ center}$ = 297 mm

tw = 62.5 mm
we = 97 mm
es = 71.5 mm
sb = 7 mm

(VI to VII)
w = 16 mm
e = 39 mm
s = 57 mm

$\sigma = 64°$
$\alpha = 59°$
$\gamma = 40°$

The meridian line is not properly graduated for either a south or a north-facing cone with gnomon point on the axis of the conical surface.

For $\varepsilon = 24°$, $\omega = 18°48'$
$\phi_W = 63°$
$\phi_S = 62°30'$

Unpublished

3074G. Conical
Herculaneum, Italy
Casa della Gemma

The gnomon in position is a restoration.

White marble with gray streaks - The area above the winter solstice has been chipped away to form a two inch indentation. Eleven hour lines extend from the winter solstice across the equinox to the summer solstice. Two legs flank a stepped base.

Provenance: Casa della Gemma, Herculaneum

H = 561 mm
W = 460 mm
D_{top} = 332 mm
D_{base} = 210 mm

tw = 70 mm
we = 71.5 mm
es = 100 mm
sb = 7 mm

(VI to VII)
w = 30 mm
e = 47 mm
s = 70.5 mm

$\alpha = 51°$
$\sigma = 63°$
$\gamma = 68°$
$\beta = 40°$

For $\varepsilon = 24°$, $\omega = 20°30'$
g = 186 mm
$\phi = 44°24'$
$\phi_w = 41°18'$
$\phi_s = 41°$

Unpublished

3075. Conical
Present Location Unknown

White marble - Eleven hour lines radiate from the gnomon hole across the equinox to the lower edge of the shadow-receiving surface.

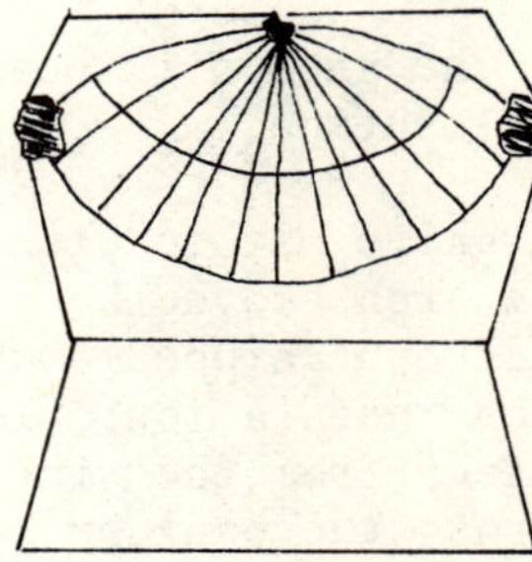

Found at Pompeii, January 22, 1762. Antonini saw dial in Naples in 1798.

Antonini, dial no. 3.
Martini, H., *Abhandlung von den Sonnenuhren der Alten*, Leipzig, 1777, p. 57, plate 1, figure 3 (drawing).

***

3076. Conical
Present Location Unknown

White marble - The thirteen engraved hour lines include two near the horizontal top edge. The fourth, seventh, and tenth hour lines are specially marked (x, CƆ , x). Three engraved day curves are visible. The oblique front is decorated with a simple geometric design which includes a crescent.

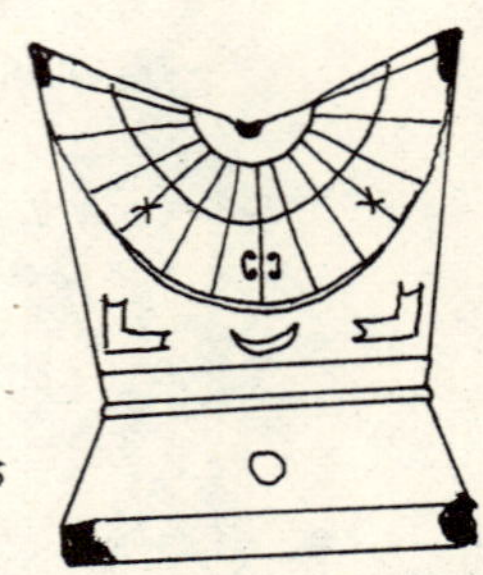

Found in the Via Appia near the Circus of Caracalla in Rome. The dial was seen in the Archiginnasio della Sapienza in Rome in 1816.

Settele, G., *Memoria sopra la forma della linee orarie indicanti le ore ineguale degli antichi sopra gli orologi solari*, Rome, 1816, p. 19.

3077G. Conical
Vatican City
Vatican Museum, Magazzino Sculture, no. 3181

White marble - The preserved hour lines (1,2,3,4,5,6,7,8) continue beyond the winter solstice toward the gnomon hole and also continue beyond the summer solstice onto the lip border. The two engraved solstices and equinox are parallel to each other. A depressed margin, stepped molding, and two narrow half columns decorate the base.

Provenance unknown.

H = 460 mm
W = 452 mm
D_{top} = 230 mm

gw = 50 mm
we = 62 mm
es = 120 mm
sb = 4 mm

(VI to VII)
w = 21 mm
e = 38 mm
s = 73 mm

$\sigma = 83.5°$
$\gamma = 44°$
$\beta = 39°$

For $\varepsilon = 24°$,
$\omega = 35°35'$
$g = 117$
$\phi = 44°$
$\phi_w = 42°48'$
$\phi_s = 46°\ 18'$

Unpublished

3078G. Conical
Vatican City
Vatican Museum, Magazzino Sculture, no. 3179

Porous stone - The much worn surface of this dial still shows eleven scratched hour lines. In taking the above measurements I have assumed certain vague traces to be the winter solstice and the equinox. Any trace of summer solstice has broken away.

Provenance unknown.

H = 142 mm
W = 238 mm
D = 85 mm

gw = 20 mm
we = 32 mm
es = 41+ mm

(VI to VII)
w = 9 mm
e = 16 mm
s = 30 mm

$\sigma = 45°$
$\gamma = 75°$
$\beta = 28°$

For $\varepsilon = 24°$, $\omega = 15°30'$
$g = 68.9$ mm
$\phi = 43°18'$
$\phi_W = 50°48'$
$\phi_S = 62°18'$

Unpublished

3079G. Conical
Vatican City
Vatican Museum, Cortile Ottagono, no. 929

A small round hole in the top surface of the dial probably held part of the gnomon.

Fine grain white marble - Seven deeply engraved hour lines (1,2,3,4, 5,6,7) are preserved in the badly damaged shadow-receiving surface. Parts of the parallel solstices and equinox remain. Two lion feet support the dial on the left and right sides of the base. The foot motif is carried around to similar reliefs on the left and right sides of the dial which also include a design which incorporates stalks springing from calyxes.

H = 453 mm
W = 445 mm
D = 275 mm

we = 60+ mm
es = 115 mm
sb = 8 mm

(I to II)
w = 19 mm
(VI to VII)
e = 37 mm
s = 64 mm

$\sigma = 58°$
$\gamma = 62°$

For $\varepsilon = 24°$, $\phi_W = 47°42'$
$\phi_S = 32°45'$

Conical dials with similar side reliefs and lion feet include 3026G, 3071G, 3064G.

Amelung, vol. 2, p. 105, table 39e (photograph).
Antonini, dial 3 (drawing).
Settele, G., *Memoria sopra la forma della linee orarie indicanti le ore ineguale degli antichi sopra gli orologi solari*, Rome, 1816, p. 18.

3080G. Conical
Rome, Italy
Museo Nazionale Romano, courtyard

White marble - The museum has restored this dial by joining two halves. Eleven hour lines cross three day curves. The third, sixth, and ninth hour lines are specially marked. A simple sloping base supports the dial face.

Provenance unknown.

For $\varepsilon = 24°$, $\phi_W = 59°12'$ and $\phi_S = 57°12'$.

Other dials in the catalogue with specially marked third, sixth, and ninth hour lines include: 1016G, 1022G, 1056G, 2023G, 3022, 3067G, 3076.

H = 339 mm
W = 327 mm
D = 255 mm

tw = 58 mm
we = 59 mm
es = 60.5 mm
sb = 4 mm
(VI to VII)
w = 16.2 mm
e = 35 mm
s = 52 mm

$\gamma = 57°$
$\beta = 41°$

Unpublished

***

3081. Conical
Present Location Unknown

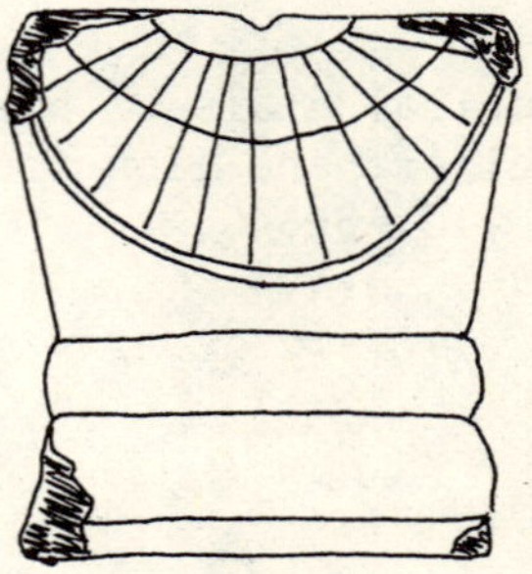

Found in the ruins of ancient Ocriculum (modern Otricoli) and placed in the Pio Clementine museum in 1787. Apparently similar to a dial from Palestrina (3084G) now in the Vatican Museum.

Guattani, G. A., *Monumenti Antichi Inediti*, vol. 2, Rome, 1787, plate 2, p. 35.
Settele, G., *Memoria sopra la forma della linee orarie indicanti le ore ineguale degli antichi sopra gli orologi solari*, Rome, 1816, p. 18.

3082. Conical
Merlara, Italy

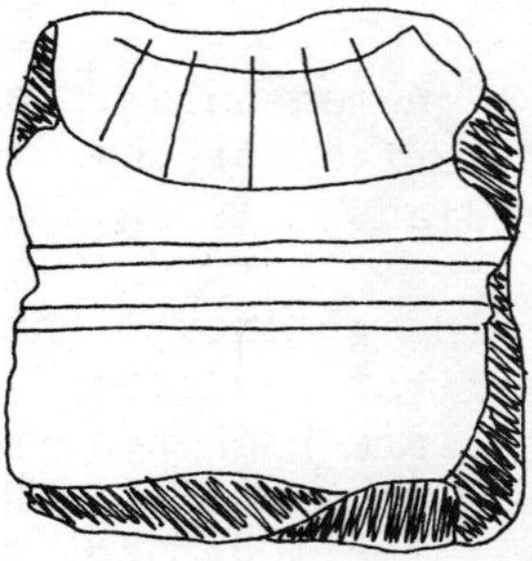

Limestone - This dial fragment appears to be conical. Parts of six hour lines (4,5,6,7,8,9) and two day curves (equinox and summer solstice) are preserved.

Found in the fields near Montagnana, Italy.

Callegari, Adolfo, "Regione X (Venetia et Histria)," *Reale Accademia Nazionale dei Lincei, Notizie degli Scavi* 18 (1940): 169, figure 2 (photograph).

***

3083G. Conical
Altino, Italy
Museo di Altino, no. AL11

H = 220 mm
W = 250 mm
D = 133 mm

gw = 30 mm
we = 55 mm
eb = 34 mm

σ = 58.5°
γ = 55°
α = 66.5°

The gnomon groove above the meridian still contains the lead fill.

White marble - Thirteen hour lines have been scratched deeply into the shadow-receiving surface from the gnomon to the lower edge. Only the lightest trace of winter solstice and equinox remain. The base exhibits a cyma profile.

Found in the necropolis of Altino.

Supposed date of origin: first century A.D.

Unpublished

3084G. Conical
Vatican City
Vatican Museum, Cortile Ottagono, no. 949

A gnomon hole is preserved in the top surface of the dial above the meridian.

Fine grained light gray marble - The first and second of eleven hour lines are broken away on this beautifully made dial. The two solstices limit the length of the remaining hour lines. Both solstices are parallel to the equinox and the lower edge. A half-cylinder molding decorates the base.

Found in Palestrina in 1786.

H = 435 mm
W = 418 mm
D = 235 mm

gw = 59 mm
we = 61 mm
es = 104 mm
sb = 10 mm

(VI to VII)
w = 25 mm
e = 40 mm
s = 65 mm

$\sigma = 74°$
$\gamma = 56°$

For $\varepsilon = 24°$, $\omega = 30°20'$
$g = 138.7$ mm
$\phi = 41°42'$
$\phi_W = 36°18'$
$\phi_S = 35°$

For other conical dials with decorative cylindrical molding see 3072G, 3067G, 3081.

Amelung, vol. 2, p. 108.
Antonini, dial 12 (drawing).

3085G. Conical
Velletri, Italy
Museo Civico

White marble - Eleven hour lines appear, at least at the equinox, to have been engraved after the three day curves. Only the meridian continues beyond the winter solstice to the gnomon hole. A simple engraved geometric design decorates the oblique front surface. Two four-toed lion feet flank the base.

H = 433 mm
W = 404 mm
D_{top} = 300 mm
tw = 47 mm
we = 60 mm
es = 98 mm
sb = 6 mm
(VI to VII)
w = 19 mm
(O to XII)
e = 35 mm
s = 59 mm
$\sigma = 72°$
$\gamma = 61°$
$\beta = 40°$

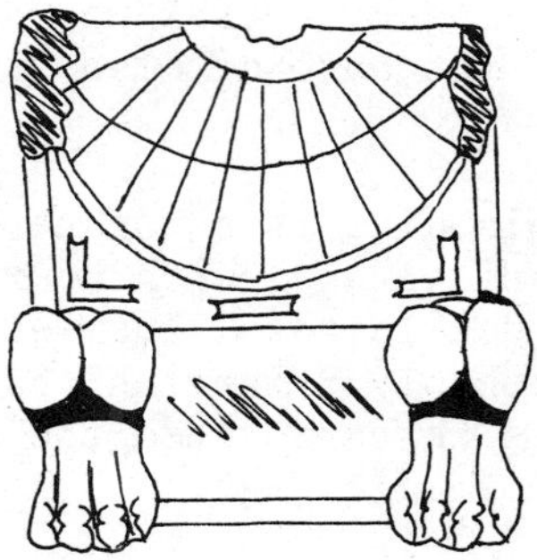

For $\varepsilon = 24°$, $\omega = 28°$
g = 130 mm
$\phi = 44°$
$\phi_w = 50°30'$
$\phi_s = 36°12'$

A similar geometric design is visible on 3072G from Pompeii and 3076 from Rome.

Cf. Antonini, dial 6 (drawing).
Wagener, A. P. and Ashby, Thomas, "Roman Remains in the Town and Territory of Velletri," AJA 17 (1913): 420.

Plate 48

3086G. Conical
London, England
British Museum, no. 1936 3-9 1

The gnomon hole has a semicircular vertical section 50 mm wide and 51 mm deep.

Stone - Eleven hour lines extend from winter to summer solstice. The three day curves have been engraved always equidistant from each other and from the lower edge of the conical surface. Dots are visible at the junctions of hour lines and winter solstice. Six shallow steps decorate the base. Greek letters have been engraved in the spaces below the equinox close to the right hour line:

Γ Δ E S Z H I

Found in 1852 near Alexandria at the base of Cleopatra's Needle.

This dial is a unique example of a conical dial with hours marked in Greek letters. The lettering is probably Byzantine, but the museum has characterized the dial as Ptolemaic.

$\phi_W = 58°$
$\phi_S = 55°$

H = 404 mm
W = 429 mm
D_{top} = 366 mm

tw = 60 mm
we = 107.5 mm
es = 105 mm
sb = 9 mm

(VI to VII)
w = 22 mm
e = 46 mm
s = 67 mm

$\sigma = 70°$
$\gamma = 53°$
$\beta = 31.5°$

A Guide to the Egyptian Collections in the British Museum, London, 1909, p. 72 (photograph) and p. 273.
Cf. Gatty, pp. 42-43.

47

48

3087. Conical
Present Location Unknown

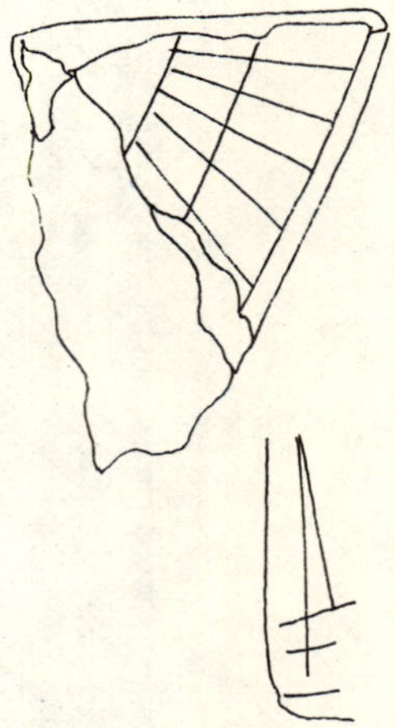

Part of a gnomon hole is visible in the top surface of the dial above the meridian.

Limestone - Only the right half of the dial surface is preserved. Five hour lines (7,8,9,10,11) and parts of three day curves are visible in the photograph. Petrie found a configuration of lines on the top surface of the fragment which he suggests were used to align the instrument.

Found in the Roman cemetery of Hawara near Memphis.

Dates from the first or second century A.D.

Petrie [2], pp. 20-21, plates 15 (photograph) and 23 (drawing).

***

3088. Conical
Carthage, Tunisia
Musée de Carthage

A gnomon hole is visible at the top of the dial.

H = 345 mm
W = 240 mm
D = 200 mm

White marble - Eleven hour lines divide the conical surface stretching from the gnomon groove to the lower edge. No day curves are visible. The base is quite plain.

Found in the ruins of Carthage.

Measurements from Ennaifer Mongi, Curator of the Musée de Carthage.

3089. Conical
Khartoum, Sudan
Archeological Museum

Subcrystalline limestone - The stone is similar to that found south of Abu Hamid, Sudan. A large portion of the dial surface is broken away, but parts of seven hour lines and the summer solstice curve are visible in Crowfoot's sketch.

H = 210 mm
W = 350 mm
D_{top} = 140 mm
D_{base} = 110 mm

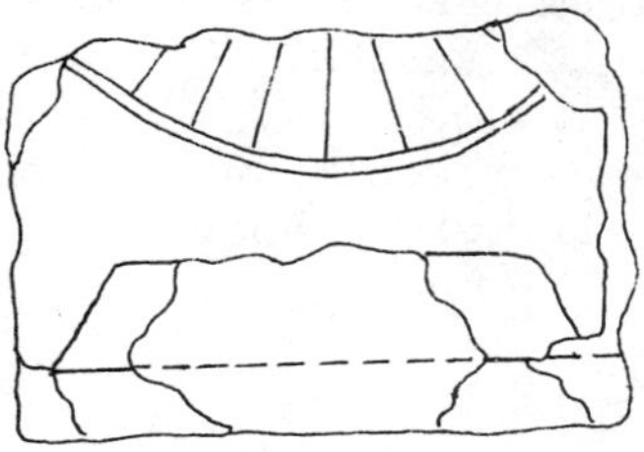

Found in the autumn of 1907 at Basa, Sudan, outside the sanctuary.

The dial is probably contemporary with the temple at Basa dating from the third century A.D. Crowfoot mentions a "similar dial" from Maryut, Egypt, preserved in the courtyard of the Alexandria Museum.

Crowfoot, J. W., "The Island of Meroë ," in *Archeological Survey of Egypt, Nineteenth Memoir*, London, 1911, pp. 17-18, plate 10 (drawing).

***

3090. Conical
Philippeville, Algeria
Musée de Philippeville

Marble - Twelve hour lines are visible in Delamare's drawing; also, two day curves can be seen. Front and sides are heavily decorated with scrolls, acanthus, and ivy.

H = 920 mm

A gnomon hole is visible in the top surface.

Gsell, S., *Exploration Scientifique de l'Algérie, Archéologie*, Paris, 1912, p. 141.
drawings by H. A. Delamare, Paris, 1850, plate 160, figures 4, 5, 6 (drawings).

3091. Conical
Rabat, Morocco
Musée Archéologique

A gnomon groove is visible in the top surface of the dial above the meridian.

W = 220 mm
tb = 160 mm

Marble - The dial surface is mounted on a column. Eleven hour lines radiate from the gnomon groove at approximately equal angles.

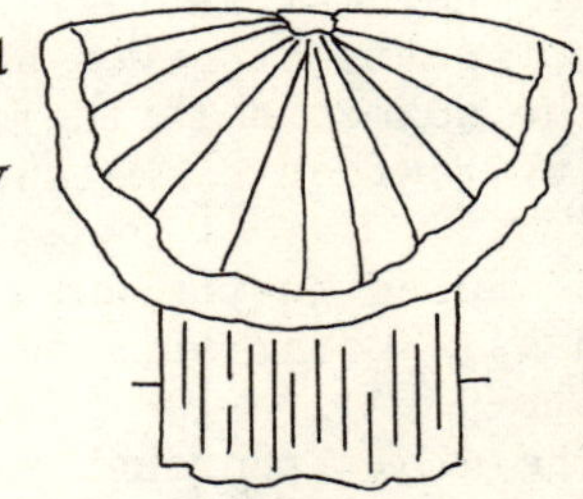

Measurements and sketch from L. Janin, Sèvres, France.

***

3092. Conical
Jerusalem, Israel
Department of Archeology, Hebrew University, no. 3018Fg616.93

Stone - Three hour lines and part of a day curve are preserved on this fragment. A drilled guilloche decorates the lip. There is a trace of red color in the lines and drilled holes.

Found next to the wall of the temple mound in an Herodian context, in Jerusalem.

***

3093. Conical
Jerusalem, Israel
Department of Archeology, Hebrew University, no. 684ßFg6

Stone - The upper left corner of the conical fragment has two hour lines and part of a summer solstice curve. An undrilled guilloche decorates the lip.

$\alpha = 64°$

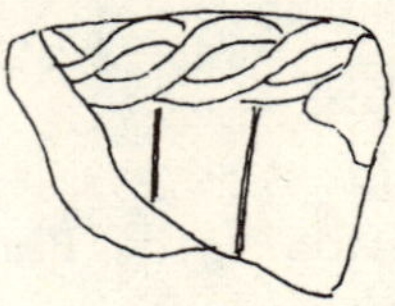

Found alond the wall of an Herodian channel, in Jerusalem.

Descriptions of 3092 and 3093 from notes of D. J. Price.

3094G. Conical
Paris, France
Louvre Museum, no. AO 4823

Plate 49
Page 310

Very hard stone - The dial consists of two fragments: AO 4823 found by Renan and preserved in the Louvre, and M 300 found by the Direction General of Antiquities of the Lebanese Republic and published by Dunand and Duru. AO 4823 has three hour lines (5,6,7) and three day curves. M 300 has five hour lines (7,8,9,10,11) and three day curves. All hour lines extend above the meridian. Both fragments contain parts of a Phoenician inscription:

[ל]אדן למלכעשתרת אלחמן אש
נד[ר] עבדך עבדאסר בן א. . .

From Oumm el'Amed, Syria.

The inscription dates from the second half of the second century B.C.

In *Mission de Phénicie*, Woepcke determines that the complete tw was equal to 83 mm for $\phi = 34°18'$ and $\varepsilon = 23°30'$.

AO 4823:

H = 410 mm
W = 230 mm
D_{max} = 220 mm

we = 71 mm
es = 128 mm

(V to VI)
w = 29 mm
e = 45 mm
s = 76 mm

$180° - \gamma = 126.5°$

Catalogue Official des Collectiones du Conservatoire des Arts et Métiers, vol. 3, 1880, no. 8280, p. 107 (photograph).
CIS, I, pp. 33-34.
Dunand, M., and Duru, R., *Oumm el'Amed*, Paris, 1962, p. 79, 110, 185 and plate 39.1.
Gatty, pp. 36-37.
Renan, E., *Mission de Phénicie*, Paris, 1874, pp. 729-44.
"Trois Inscriptions Phéniciennes trouvées à Oumm el-Awamid," JA 20(1862): 378-80.
Woepcke, F., "Note de M. Woepcke sur le Cadran Solaire Phénicien de M. Renan," JA 1(1863): 292-94.

49

3095. Conical
Mada'in Salih, Saudi Arabia

Sandstone - Eleven hour lines divide two day curves and the lower edge. The base bears a Nabatean inscription dating from the first century A.D.

H = 420 mm
W = 350 mm
D = 250 mm

מנשא בר נתן שלם

Janssen, A. J., and Savignao, R., *Mission Archéologique en Arabie*, vol. 1, Paris, 1909, pp. 302-04, figure 113 (drawing).
Lidzbarski, M., *Ephemeris für Semitische Epigraphik*, Giessen, 1909-15, pp. 87-88.

***

3096. Conical
Palmyra, Syria

A photograph from Ingholt shows that this dial is part of a dedicatory column. Nine hour lines and three day curves are visible. A bilingual inscription in Greek and Palmyrenian, dating from 64 A.D. also appears on the column drum:

H = 630 mm
W = 445 mm

.....ουσατ...
......τας•ν...
Σαλαμ]άλλαθον•[Ἰαριβωλέους τοῦ
Νουρ]βήλου•τοῦ•ἐπικαλ[ουμένο]υ..α.ι.
Παλ]μυρηνὸν•φυλῆς•τῆς•αὐτῆς•εὐσεβῆ•καὶ
ἀρέ]σαντα•αὐτοῖς•ἐν•ῶολλοῖς•ῶράγμασιν
ἐ]ν τε•κτίσμασιν καὶ•ἀναθήμασιν
ἀναλώμασι•τε•οὐκ ὀλίγοις•τειμῆς
ἕνεκεν (ἔ)το(υ)ς•εοτ´μηνὸς•Περειτίου.

צל[מא] דנה די שלמלת בר
[ירחבו]לא בר נורבל די מתקרה
א[. .]ר ידיעבל חדמריא די מן

3096. continued

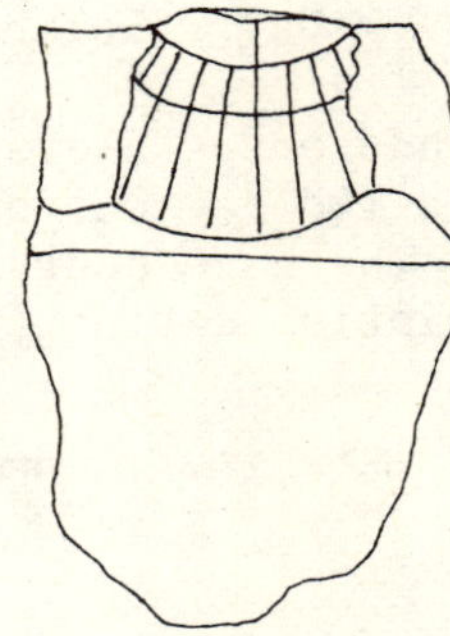

בנ[י מ]עזין די אקימו לה אלת

[. . . . . . ב]דיל די קם ובנא וקרב

[. . . . . . ושפר ל]הן בכל גנס כלה

[. . . . . צ]למא דנה ועמודא

[דנה בירח שבט] שנת 37[5]

Cantineau, J., *Inventaire des Inscriptions de Palmyre II*, Beyrouth, 1930, no. 1, pp. 22-23 (photograph of inscription).
CIS, II, 3966.
Wiegand, T., and Krencker, D., *Palmyra*, Berlin, 1932, pp. 89-91, 153, 160 (photograph and drawing).

***

3097. Conical
Aix-les-Bains, France
Archeological Museum

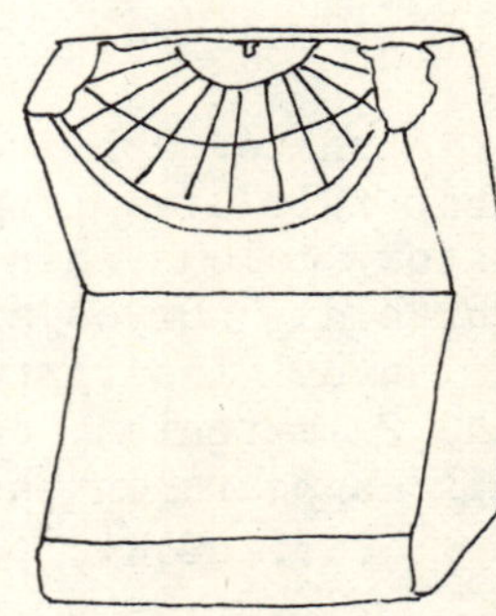

White marble - This simple conical dial was found near the ruins of the Roman baths at Aix and probably dates from the late empire.

Sketch and description from L. Janin, Sèvres, France.
Cf. Rohr, René, "Le Cadran Solaire Romain de Bettwiller," *Cahiers Alsaciens d'Archéologie d'Art et d'Histoire*, Strasbourg, 1966, pp. 50, 54.

***

3098. Conical
Auriol, France

Renan reports that he has a drawing of a dial similar to 3094G, but which has a Latin inscription.

Renan, E., "Trois Inscriptions Phéniciennes trouvées à Oum el-Awamid," JA 20(1862): 379, note 1.

3099. Conical
Auxterre, France
Musée Archéologique

Common stone - Only the left half of the conical shadow-receiving surface is preserved. Six hour lines are visible uncrossed by day curves. The dial is supported by the head of a young woman.

H_{head} = 120 mm

For other dials with hour lines only see: 3003G, 3017G, 3018G, 3044G, 3066G, 3088, 3091, 3103.

Espérandieu, E., *Recueil Général des Bas Reliefs, Statues, et Bustes de la Gaule Romaine*, vol. 4, Paris, 1907, p. 94, no. 2924 (photograph).

***

3100G. Conical
Avenches, Switzerland
Musée Romain, no. 62:93

Limestone - The back of this dial is cylindrical, like a column. One day curve crosses the hour lines (1,2,3,4,5,6,7,8) near the top edge. The hour lines continue lightly above this day curve to the top center of the shadow-receiving surface. Divisions on the day curve and the lower edge are equal except between the top edge and the first and last hour lines.

Found in 1962 in a private house in Avenches (ancient Aventicum).

H = 322 mm
W = 190 mm
D = 250 mm

tb = 185 mm
tc = 35 mm
cb = 150 mm

(V to VI)
c = 8 mm
b = 36 mm

$\alpha = 46°$

Unpublished

3101. Conical
Kassel, Germany
Hessisches Landesmuseum

A gnomon hole is visible in the top surface above the meridian.

H = 240 mm
W = 250 mm

White marble - Eleven hour lines divide three day curves. Two four-toed lion feet flank the base.

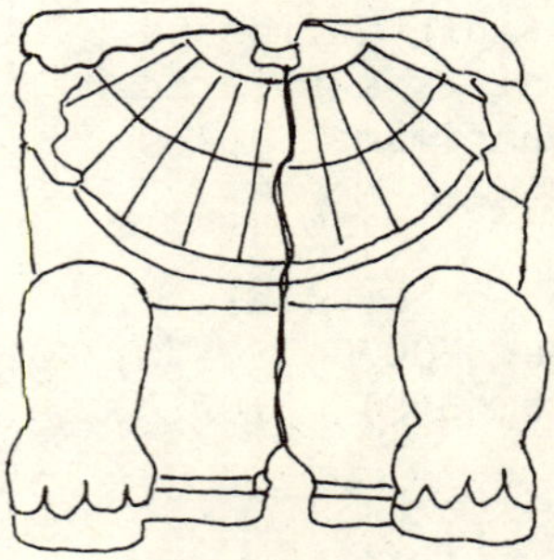

Provenance unknown.

The dial is of Roman origin, dating from the first century A.D.

Michel, H., *Scientific Instruments in Art and History*, New York: Viking Press, 1967, p. 159 and plate 65 (photograph).

***

3102. Conical
Luxemburg, Luxemburg
Les Musées d'Etat, Halle 13, Case 13

A gnomon of dubious origin is visible in the gnomon hole.

H = 250 mm
W = 400 mm

Limestone - Eight hour lines are preserved on the western half of this sundial fragment. Each line is emphasized with red paint, and all lines converge to the gnomon hole. No day curves can be seen.

Information taken from a photograph from Mr. and Mrs. R. S. Webster.

3103. Conical
Saint-Bertrand de Comminges, France
Archeological Museum

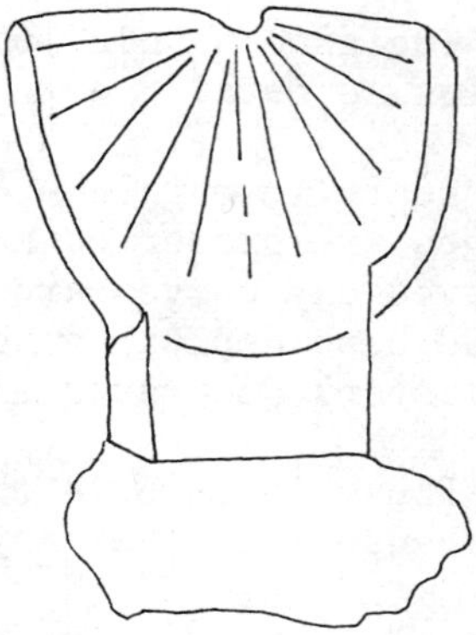

Limestone - This conical dial with hour lines only was found in the street of the western sector of Saint-Bertrand in 1929. Money found nearby dates from the first century of the late empire.

For other dials with hour lines only see: 3003G, 3017G, 3018G, 3044G, 3066G, 3088, 3091, 3099.

Lavedan, P., Lizop, R., Sapene, B. *Les Fouilles de Saint-Bertrand de Comminges*, Toulouse, 1929, p. 41 and plate 17, figure 2 (photograph).

3104G. Conical
Vienna, Austria
Kunsthistorisches Museum, no. 1736

The gnomon hole, 60 mm deep, contains no metal.

Pentelic marble - All eleven hour lines are preserved and three engraved day curves are visible. A wide band depicting amphorae between pilasters decorates sides and back.

Found in 1896 in the region of Siebengebirge, Germany.

$H = 300$ mm
$W = 450$ mm
$D_{top} = 340$ mm

$tw = 83$ mm
$we = 60$ mm
$es = 76$ mm
$sb = 19$ mm

(VI to VII)
$w = 27.5$ mm
$e = 40$ mm
$s = 55$ mm

$\sigma = 51.5°$
$\beta = 42.5°$

For $\varepsilon = 24°$, $\omega = 14°30'$
$g = 181$ mm
$\phi = 40°12'$
$\phi_W = 38°36'$
$\phi_S = 36°24'$

Computed parameters, material, and workmanship suggest Greek origin--possibly Athens.

Unpublished

3105G. Conical Fragment
London, England
British Museum, no. 55 10-30 30 CN

Gray streaked marble - This fragment is the upper right corner of a conical dial. Three hour lines and three day curves are preserved. All engraved lines are very fine. The back of the fragment is curved. An engraved line on the top surface of the fragment coincides with the equinox.

H = 153 mm
W = 132 mm
D = 70 mm

sb = 2 mm

(X to XI)
w = 17.2 mm
e = 27 mm
s = 43 mm

$\alpha = 52.5°$

Unpublished

3106G. Conical Variant — Plate 50, Page 325
Rhodes, Greece
Archeological Museum, study collection

H = 350 mm
W = 370 mm
D = 120 mm

we = 48 mm
es = 98 mm

(VI to VII)
w = 18 mm
e = 29.7 mm
s = 55 mm

σ = 78°
γ = 56°

At the top left corner of the dial, above the left meridian, there is a gnomon hole containing some trace of lead fill.

Pentelic marble - The shadow-receiving surface of this dial is essentially that of a section of truncated cone which has been split at the meridian and transposed before being cut into the stone. Rather than continuing smoothly to the top of the dial, the conical surface is cut off above the winter solstice by a rough protrusion on each side. Six hour lines, including the meridian, are crossed by three day curves within each dial half. The summer solstice is distinct from the lower edge, but the winter curve is marked only by the junction of the smooth conical surface with the rough protrusion. The center front surface of the dial has been decorated with a relief showing a priestly figure wearing a *polos* and a flowing robe, reaching for the tendrils of a symmetrically framing plant.

Found in 1968 in the ruins of a Paleochristian basilica on Rhodes.

A remarkably similar decorative design, dating from the fourth century B.C., appears on the back of a marble chair from the acropolis of Athens.

For $\varepsilon = 24°$, $\omega = 37°30'$, g = 62.6 mm, and $\phi_W = 33°$, with $\phi_S = 37°$.

Archaiologikon Deltion, Chronika 24(1969): 466 and plate 471 (photograph).
Cf. Kraus, T., *Die Ranken der Ara Pacis*, Berlin, 1953, p. 53 and plate 12 (photograph of chair).

3107G. Conical Variant
Sparta, Greece
Archeological Museum, no. 6193

H = 279 mm
W = 410 mm
D = 500 mm

Two square holes diametrically opposite from each other in the hexagonally shaped top surface of the dial are probably gnomon holes. Gnomons placed in these holes could have served all four shadow-receiving surfaces. The central rectangular hole (132 mm long, 20 mm wide, and 56.5 mm deep) and associated grooves still contain a trace of lead and may have been part of the gnomon system. More probably they held some ornament or additional timekeeping device.

White marble - The dial is supported by a ship-shaped base decorated on the left and right sides by reliefs depicting bearded giants whose tails end in acanthus leaves. Four separate line grids have been engraved on the transposed halves of the north and south facing conical surfaces.

A dial with a similar ship base, but without reliefs was found in the gymnasium on Delos (see 1072G).

Basch observes that the ship is characteristic of those of the Hellenistic period. The Triton figure dates from the second century A.D., possibly copying a Hellenistic design.

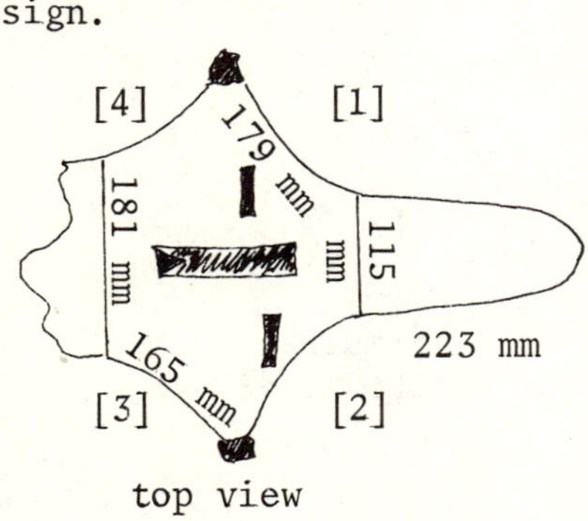

top view

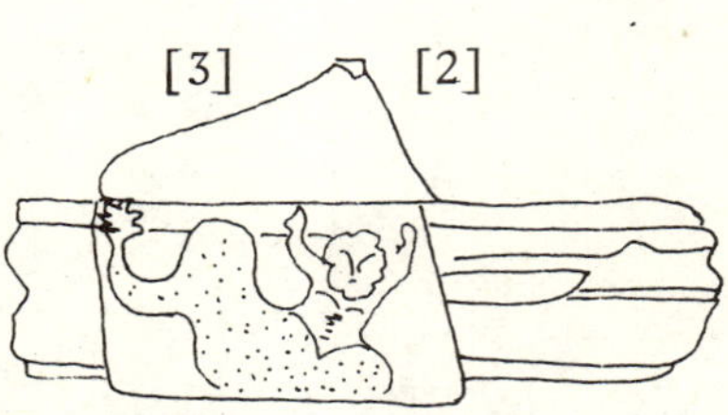

west side

3107G. continued

South Face (Surfaces 1 and 2)

Hour lines on this face extend from the engraved winter solstice across the engraved equinox and summer solstice to the lower edge. The meridian is not engraved on either surface, but ten other hour lines are preserved.

For $\varepsilon = 24°$, $\omega = 44°16'$ and $\phi = 38°46'$, with $\phi_W = 36°12'$ and $\phi_S = 27°50'$.

tw = 30 mm
we = 30 mm
es = 76 mm
sb = 5 mm

(V to VI)
w = 11 mm
(VI to VII)
e = 20 mm
s = 30 mm

$\sigma = 79.5°$
$\gamma = 33.8°$
$\alpha = 56.5°$

North Face (Surfaces 3 and 4)

Measurements were taken from the better preserved west side segment. On it, five hour lines (7,8,9,10,11), winter solstice and equinox are scratched. the seemingly poor engraving job may only be the effect of weathering. The meridian probably coincides with the west edge. It is possible that the lower edge marked the summer solstice. Measurements taken on the east side confirm those on the west.

tw = 54 mm
we = 30 mm
eb = 78 mm

(X to XI)
w = 11 mm
(VII to VIII)
e = 16 mm
b = 23 mm

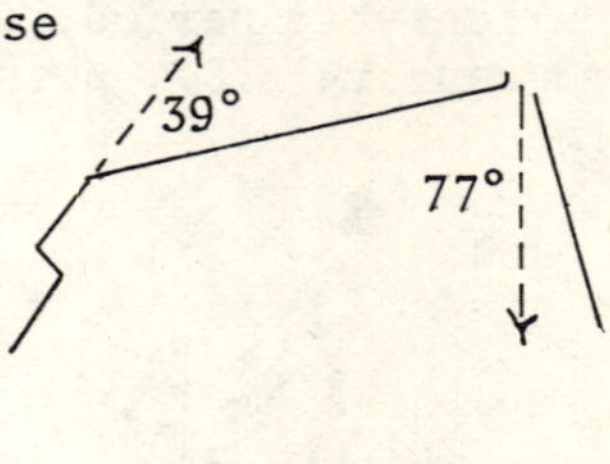

Basch, L.,"Trois modèles de navires in marbre au Musée de Sparte," *Antiquité Classique* 38(1969): 430-46.

3108G. Conical Variant
Cambridge, England
Fitzwilliam Museum, no. GR 100-1906

H = 464 mm
W = 395 mm
D = 228 mm

σ = 62.5°
γ = 43.2°

left:
tc_1 = 67 mm
c_1c_2 = 16 mm
c_2c_3 = 63 mm
c_3c_4 = 139 mm
c_4b = 5 mm

(VI to VII)
c_1 = 18.2 mm
c_2 = 19 mm
c_3 = 31.2 mm
c_4 = 57 mm

right:
tc_1 = 61 mm
c_1c_2 = 16 mm
c_2c_3 = 64 mm
c_3c_4 = 141 mm
c_4b = 10 mm

(V to VI)
c_1 = 16.5 mm
c_2 = 19.8 mm
c_3 = 31.2 mm
c_4 = 55.5 mm

The gnomon hole at the left side is 8 mm in diameter. It slants downward at an angle 45° from the horizontal. The right gnomon hole is lead filled.

White Egyptian limestone - The transposed halves of a dished conical surface form the shadow-receiving area of this dial. The sagitta at the left equinox is about 8 mm; at right, 9 mm. Six hour lines, including the meridian, are preserved on each side. Only the meridian continues above the winter solstice to the gnomon hole. Four day curves, engraved always equidistant from each other and from the lower edge of the dial surface, are visible. The base is a schematic lion's paw.

The dial has been characterized as an Egyptian work of Imperial date.

Other dials with transposed halves include: 1072G, 1073, 1074, 3106G, 3107G, 5002G, 7007.

Budde, L., and Nicholls, R., *A Catalogue of the Greek and Roman Sculpture in the Fitzwilliam Museum, Cambridge*, Cambridge, 1964, pp. 112-13, no. 185 and plate 60 (photograph).

3109G. Conical Variant
Izmir, Turkey
Archeological Museum, Basmahane, no. 6886

Plate 51
Page 325

White marble - The unique shape of this dial face is shown in the adjoining meridian section. The surface is conical, possibly a complete right circular cone. The vertex of the conical surface is carved out of the stone. Its position is marked by a shallow, square, depression which contains no metal. The generator of the conical surface is horizontal at the meridian. Seven preserved hour lines, containing traces of black coloring, cross three day curves, but do not extend beyond either solstice. The day curves do not seem to be parallel to each other, and certainly the summer solstice is not always equidistant from the lower edge.

H = 381 mm
W = 320 mm
D_{base} = 160 mm

vw = 51 mm
we = 53 mm
es = 100 mm

(VI to VII)
w = 19.5 mm
e = 29 mm
s = 52.5 mm

$\beta = 36.5°$
$\gamma = 52.5°$

$\phi = 37°30'$
$\varepsilon = 24°$

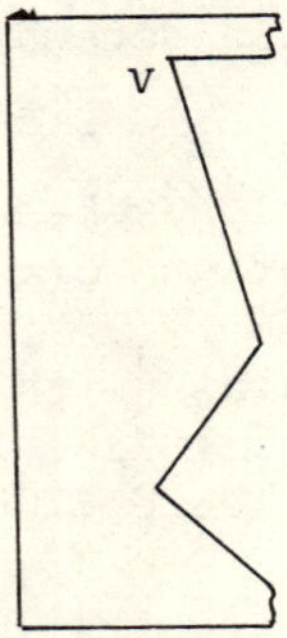

All roofed spherical sundials have catalogue numbers beginning with 2.

Unpublished

Horizontal Dials

Measured Quantities

H = height
W = width
D = depth
d = diameter of wind circle

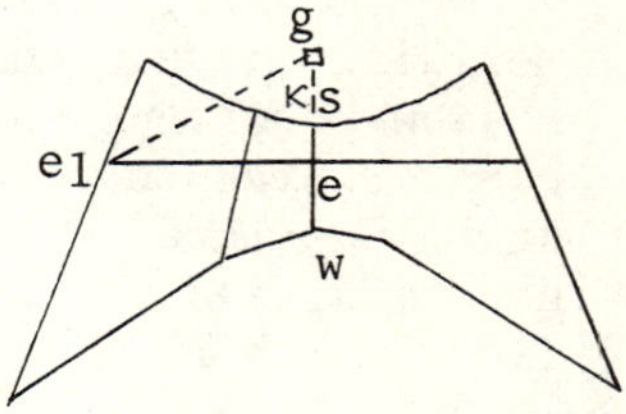

gs = distance along the meridian from the center of the gnomon hole to the summer solstice curve
se = distance along the meridian from the summer solstice curve to the equinox curve
ew = distance along the meridian from the equinox curve to the winter solstice curve
$ge_{1(2,3,4,5)}$ = distance between the center of the gnomon hole and the first (2,3,4,5) hour point on the equinox
$\kappa(\mu,\nu,o,\pi)$ = angle between the noon line and the line connecting the center of the gnomon hole with the first (2,3,4,5) hour point on the equinox

Calculated Quantities

ϕ = intended latitude
ε = obliquity

h = perpendicular height of gnomon
fs = noon shadow at summer solstice relative to the projection of the gnomon point on the dial face

A_3 = direction angle for the third seasonal hour at the equinox ($\tau = 45°$)
Q_3 = shadow length for the third seasonal hour at the equinox relative to the center of the gnomon hole

4001G. Horizontal Plane — Plate 52
Delos, Greece
Archeological Museum, study collection

H = 366 mm
W = 503 mm
D = 57 mm

gs = 30 mm
se = 23 mm
ew = 49 mm

ge_1 = 220.5 mm
ge_2 = 115 mm
ge_3 = 80.5 mm
ge_4 = 64 mm
ge_5 = 57 mm
κ = 75°
μ = 61°
ν = 47°
ο = 31.5°
π = 15.5°

Traces of iron gnomon remain in gnomon hole 13 mm in diameter.

White marble streaked with gray - Meridian and equinox line divide the shadow-receiving surface into four equal rectangles. The usual eleven hour line/three day curve grid has been modified to include two straight lines which radiate from the point of intersection of meridian and winter solstice at angles of 25° (left) and 24° (right). A Greek inscription along the left line above the summer solstice reads

ποῦ χρόνος πάσης ἡμέρας παρήκει

along the right line

ποῦ χρόνος πάσης ἡμέρας λοιπός

along the summer solstice

τροπαὶ θ[ερι]ναί

along the equinox

ἰσημερία

along the winter solstice

τροπ[α]ὶ χ[ε]ι[με]ριναί

Found in a house during the excavations of Couve in 1894.

For ε = 24°, φ = 39°,
h = 42 mm
fs = 11 mm

Other dials with engraved daylight triangles include 3046, 1068G, 1044.

Couve, Louis, "Fouilles à Délos," BCH 19(1895): 476, no. 1.
Deonna, pp. 193-94, figure 230 (drawing).
cf. Reinach, T., "Fouilles de Délos," BCH 29(1905): 251.

50

51

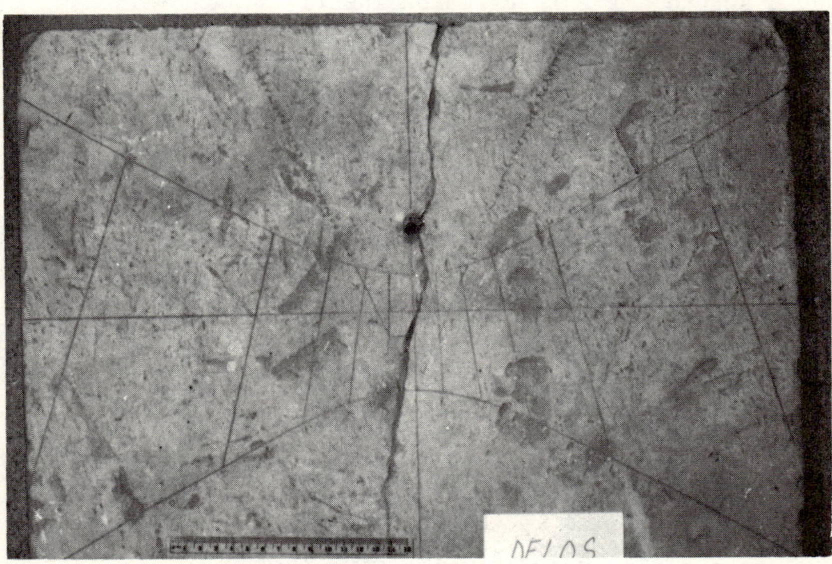

52

4002G. Horizontal Plane
Aquileia, Italy
Archeological Museum, garden

Plate 53
Page 330

table
H = 64 cm
W = 1 M
D = 2.06 M

rose
d_i = 614 mm
d_o = 660 mm
gs = 37.2 mm
se = 30 mm
ew = 74 mm

The iron gnomon, now broken off at the base, is held in place by a lead adhesive.

Limestone from Karste in Yugoslavia - A permanent glass covering makes modern measurement impossible. The figures at left were obtained from Grablovitz' early publication and recent photographs. These researched and calculated results compare favorably with measurements taken from a plaster cast of the dial displayed in the Museo della Civiltá Romana. The windrose surrounding the dial is inscribed in Latin

EVRVS AVSTER AFRICUS FAONIVS
AQUILO SEPTENTRIO BOREAS DESOLINVS

and a second Latin inscription identifies the maker

M ANTISTIVS EVPORVS FECIT

Both inscriptions, the eleven hour lines and three day curves are emphasized by a black coloring which may not be original.

The dial was found in 1878 near the ruins of a small Jupiter temple in the northwestern part of the Roman city of Aquileia.

Kenner dates the epigraphy to the republic or early empire.

Drecker computes the following parameters:
$\phi = 43°32'$, $\varepsilon = 24°$, h = 50.4 mm, fs = 17.9 mm

For other dials inscribed with wind names, see 4008G, 4010, 4009, 5001.

4002G. continued

Grablovitz, G., "Sull'Orologio Solare Scolpito nel Monumento Scoperto il Novembre, 1878, nel Fondo Cassis alle Marignane d'Aquileia," *Archeografo Triestino* 13 (1887): 209-25.
Kenner, pp. 7-18, figures 7, 8, 9, 10 (drawings).
Cf. Bracchi, col. 62-65 (drawings).
Brusin, G., *Il R. Museo Archeologico di Aquileia*, Rome, 1936, p. 33, figure 94 (photograph).
Diels, pp. 182-84, figures 60, 61 (drawings).
Drecker, pp. 46-47.
Gatty, p. 44.
Gregorutti, *Bulletino*, 1879, pp. 28-30 (drawing).
Rehm, A., "Griechische Windrosen," *Sitzungsberichte der Königlich Bayerischen Akademie der Wissenschaften (Philosophisch-philologische und historische Klasse)*, 1916, p. 68, fig. 10 (drawing).
Weiss, E., "Horizontal-Uhren," *Mitteilungen der K. K. Centralcommission zur Erforschung und Erhaltung der Kunst und Historischen Denkmale* 6 (1880): 21-23.

***

4003G. Horizontal Plane
Aquileia, Italy
Archeological Museum, study collection

H = 235 mm
W = 192 mm
D = 57 mm

White limestone - Traces of three hour lines, equinox, and one solstice remain on this dial fragment. It is possible that the bottom edge follows the curve of the other solstice. A red 392 has been painted on the back of the dial.

Unpublished

4004G. Horizontal Plane
Pompeii, Italy
Granario

The base of a bronze gnomon, 12 mm wide, remains in a lead-filled hole.

White marble - Two round holes at the left side of the dial block may have served for mounting. Meridian and equinox divide the shadow-receiving surface into four rectangles. The grid of eleven hour lines and three day curves may have been determined by the "dots" which mark the junctions of hour lines and solstices.

H = 360 mm
W = 231 mm
D = 92 mm

gs = 45.5 mm
se = 16.5 mm
ew = 38 mm

ge_1 = 145.5 mm
ge_2 = 79.5 mm
ge_3 = 62 mm
ge_4 = 52.5 mm
ge_5 = 48.5 mm

κ = 71°
μ = 53.9°
ν = 40°
o = 25.6°
π = 13.2°

For ε = 24°, ϕ = 41°54'
h = 29 mm
fs = 9 mm
A_3 = 40°18'
Q_3 = 56.7 mm

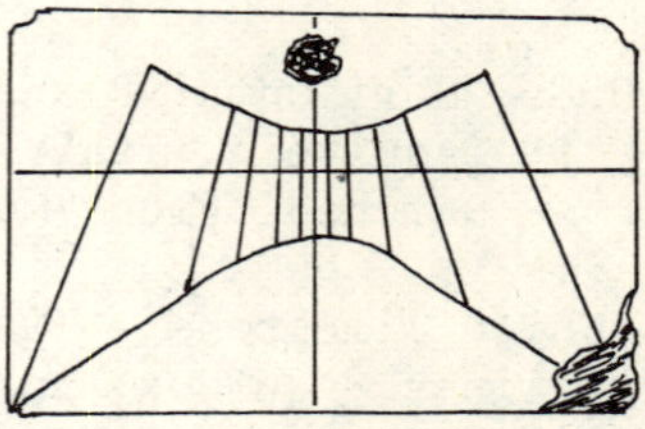

Gabba, J. and Jacono, L., "Meridiana." *Enciclopedia Italiana,* Rome, 1935-43, p. 903 (photograph, second from left).

4005G. Horizontal Plane
Pompeii, Italy
House of the Figured Capitals

Plate 54
Page 330

The core of the gnomon hole, containing no metal, is visible in the center of a large circular scarred area along the meridian above the summer solstice.

Cipolin marble, white and green - This dial has been carved on the horizontal section of a column 43 cm in diameter. The grid of lines includes eleven hour lines and three day curves. The points of intersection of hour lines and solstices are marked by "dots." The meridian extends beyond winter and summer solstices to the edges of the section.

Found *in situ* in 1832.

The House of the Figured Capitals dates from the Samnite period (200-80 B.C.).

For $\varepsilon = 24°$, $\phi = 50°36'$
$h = 24$ mm
$fs = 12$ mm
$A_3 = 26°$
$Q_3 = 67$ mm

Compare 4004G, also from Pompeii.

column
H = 193 mm
D = 430 mm

gs = 34 cm
se = 17.5 mm
ew = 59 mm

$ge_1 = 125.5$ mm
$ge_2 = 72.2$ mm
$ge_3 = 61$ mm
$ge_4 = 54$ mm
$ge_5 = 52$ mm

$\kappa = 67.5°$
$\mu = 46.8°$
$\nu = 34.9°$
$o = 21.8°$
$\pi = 11.6°$

Avellino, F. M., *Descrizione di una Casa Pompeiana.* Naples, 1837, pp. 29, 60, and table 3, no.5 (drawing).
Cf. Bilfinger, G., *Die Zeitmesser der Antiken Völker.* Stuttgart, 1886, p. 36 (drawing).
Diels, p. 181, figure 59 (drawing).

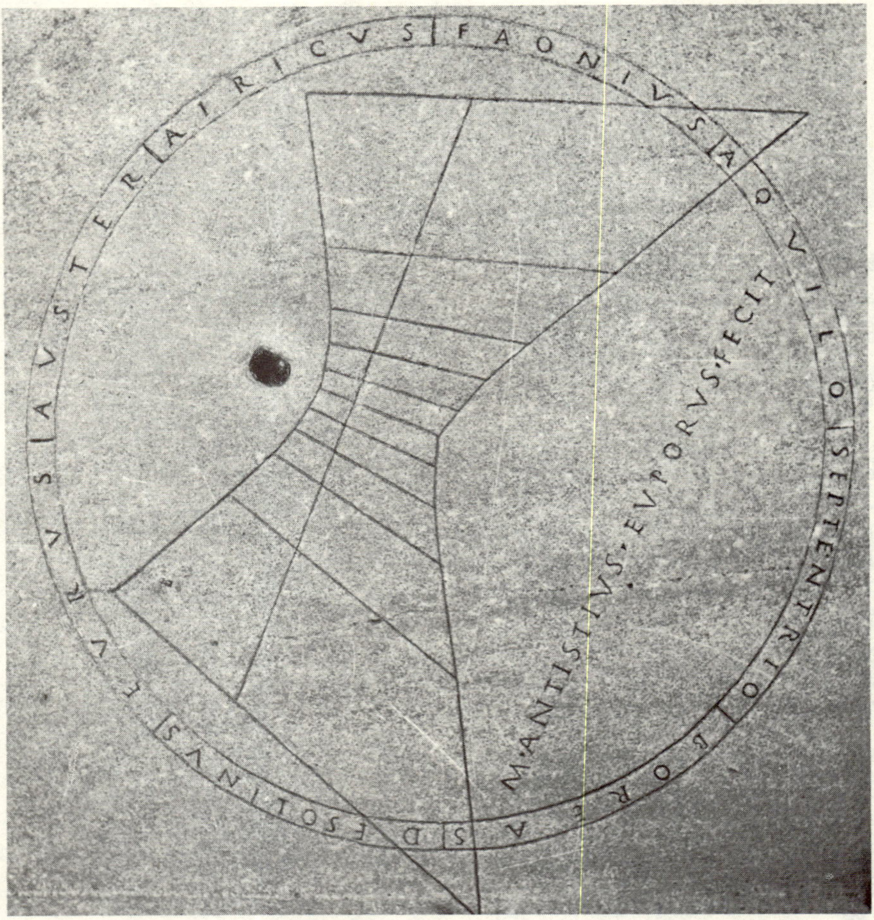
AFRICVS
FAONIVS
AQVILO
SEPTENTRIO
BOREAS
EVRVS
AVSTER
M·ANTISTIVS·EVPORVS·FECIT

53

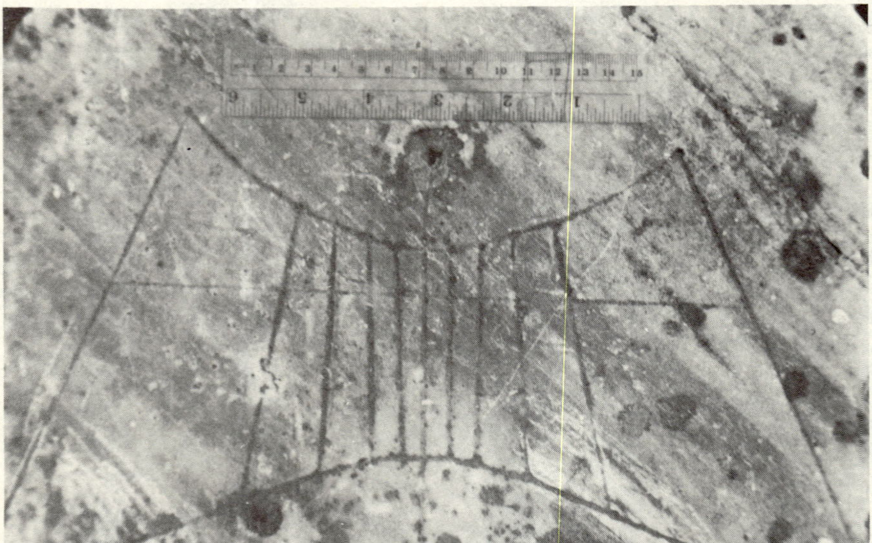

54

4006. Horizontal Plane
Pompeii, Italy?

The dial is pictured fourth from left in a photograph of "Sundials at Pompeii". Another planar dial in the same photograph (second from left) can be identified with 4005G.

Gabba, J., and Jacono, L., "Meridiana," in *Enciclopedia Italiana*, Rome, 1935-43, p. 903 (photograph).

***

4007. Horizontal Plane
Naples, Italy?

H = 530 mm
W = 345 mm

Found at Pompeii in 1865 and deposited in the museum in Naples. I was unable to locate it in the spring of 1971. The dial appears to consist of eleven hour lines, including an extended inscribed meridian, and seven inscribed day curves. The inscription along the meridian reads:

μεσηνβρία

The summer solstice is inscribed:

τροπ(ικὸs) θερ(ινόs).

The equinox is inscribed:

ἰσημερ(ινόs).

The winter solstice is inscribed:

τροπ(ικὸs) χειμ(ερινόs).

The spaces between the day curves are inscribed with the signs of the zodiac:

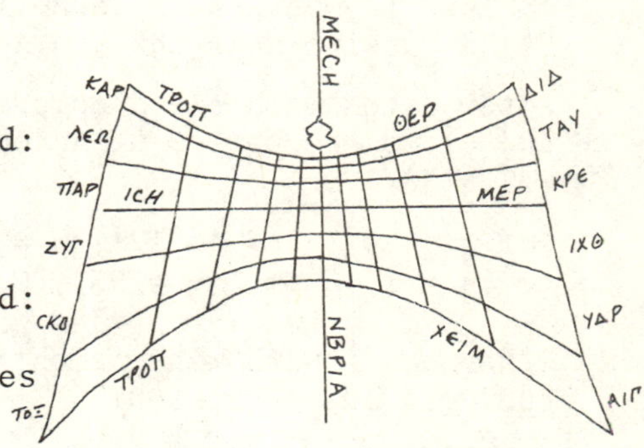

καρ(κίνοs)	δίδ(υμοι)
λέω(ν)	ταῦ(ροs)
παρ(θένοs)	κρε(ιόs)
ζυγ(όs)	ἰχθ(ύεs)
σκο(ρπίοs)	ὑδρ(οχόοs)
τοξ(ότηs)	αἰγ(όκερωs)

Fiorelli, G., *Giornale degli Scavi di Pompei*, vol. 3, Naples, 1865, pp. 14-16 with drawing (IG,XIV,705 with drawing).
Cf. Gatty, p. 46.

4008G. Horizontal Plane — Plate 55
Vatican City — Page 352
Vatican Museum, collection of Lateran Museum 10446

A trace of gnomon hole is preserved along the broken right edge which coincides with the meridian of the dial. No metal remains.

Cipollin marble, white - Three fragments have been joined to form the left half of a sundial. The equinox crosses and extends a little beyond the six preserved hour lines. Winter and summer solstice curves have been approximated by straight lines. The meridian extends beyond both solstices the length of the diameter of the inner circle. The windrose is inscribed in Greek:

[νότ]ος εὐρονοτος εὖρος ἀπηλιώτης καικίας

and the left end of a line marking the intersection of the first vertical plane and the dial plane reads:

ἰσημε(ρινή)

The fragments were found in 1814 on the Vigna Cassini near the Via Appia, Rome.

Peter dates marble and epigraphy (lunate c and ϵ) to the period of Antoninus Pius (ca. 138 A.D.).

For $\varepsilon = 24°$, $\phi = 43°12'$, $h = 39$ mm
$fs = 13.5$ mm, $A_3 = 55°$, $Q_3 = 65$ mm

Other dials with wind names include 4002G, 4010, 4009, 5001.

H = 540 mm
W = 350 mm
D = 360 mm

rose
$d_i = 604$ mm
$d_o = 666$ mm

gs = 15 mm
se = 23 mm
ew = 56 mm

$ge_1 = 207$ mm
$ge_2 = 102$ mm
$ge_3 = 67$ mm
$ge_4 = 48$ mm
$ge_5 = 40$ mm

$\kappa = 79°$
$\mu = 69°$
$\nu = 56°$
$o = 43°$
$\pi = 25°$

IG, XIV, 1308 (drawing).
Lais, P. G., *Pubblicazioni della Specola Vaticana*, vol. 4, Rome, 1894, book 1 (photograph).
Peter, F. R., *Di un Antico Orologio Solare*, Rome, 1815 (drawing).
Cf. Diels, p. 182, table 15,2 (drawing).
Gatty, pp. 45-46.
Kenner, p. 12.

4009. Horizontal Plane
Rome, Italy
Museo Nazionale Romano, nos. 40621 and 40642

40621		
H_{max}	=	220 mm
W_{max}	=	225 mm
D	=	38 mm
40642		
H_{max}	=	325 mm
W_{max}	=	210 mm
D	=	32 mm

Both these marble fragments seem to be parts of the same dial although the two pieces are of different thicknesses. The inscribed wind rose reads:

[ἀπαρκία]s βορέας καικίας ἀφηλιώτ[ηs] ...[λιβ]όνοτοs λ[ίψ].

The solstice and equinox curves are inscribed as follows:

[ὧραι θερι]ναί
[ὧραι ἰσημερ]ιναί
ὧρα[ι χειμ]ερ[ιναί].

40642

40621

Moretti, L., "Iscrizioni Greche Inedite di Roma," *Archeologia Classica* 8(1956): 69-79 (drawings).

***

4010. Horizontal Plane
Present Location Unknown

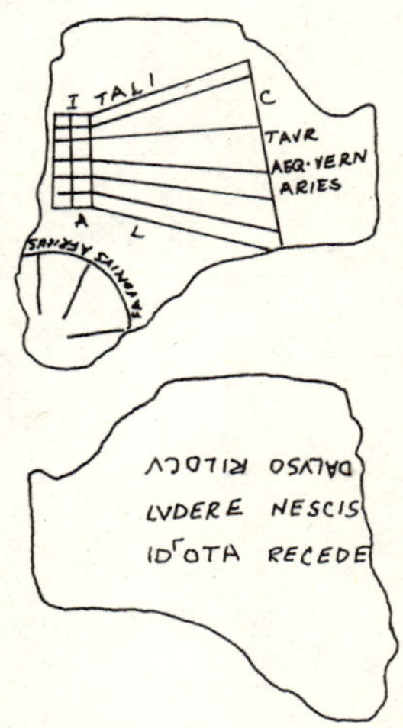

Marble - One side of this flat stone is inscribed with part of a sundial and wind rose. The solstices, equinox, and signs of the zodiac are identified in Latin:

[SOLST]ITALI[S]
AEQ[UINOCTIALIS]•VERN[ALIS]
[BRUM]AL[IS]
G[EMINI]
TAVR[US]
ARIES

The Latin names of two winds can be seen:

FAVONIVS AFRICVS

The reverse side of the stone is inscribed as a gameboard.

Found in February, 1883, in the Mausoleum of Augustus, Rome.

Atti della Reale Accademia dei Lincei 11(1883): 127 (drawing).

4011. Horizontal Plane
Present location unknown

Found near Tivoli, Italy, during tunneling for diversion of the river Aniene in 1832.

H = 3 palmi = 220 mm
W = 2½ palmi = 184 mm
D = 1 palmi = 73 mm

A Latin inscription above and below dial grid reads:

T·HERENNIVS·L·F
T·HERENNIVS·T·F
IIII·VIR·IVR·DIC·RES(TIT)

Seen by Mommsen in the Villa Gregoriana in 1847.

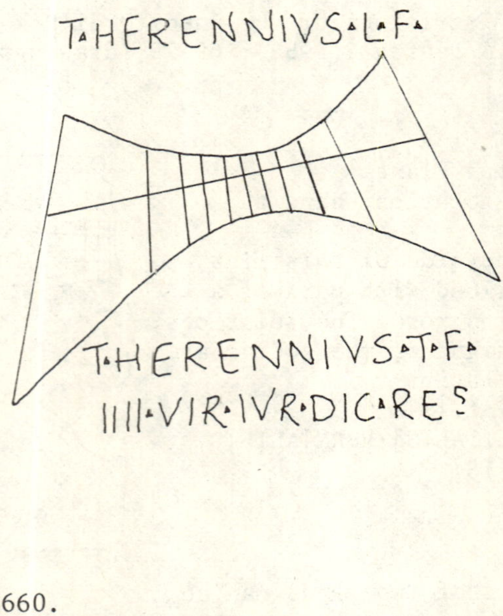

CIL, XIV, 3660.
Rossi, S. and Settele, G., "Sopra vari monumenti, e particularmente di un antico orologio solare, trovati in Tivoli in occasione degli sterramenti fatti per la deviazione dell'Aniene," *Bullettino del'Instituto di Corrispondenza Archeologica per l'Anno 1838*, Rome, 1838, pp. 97-109 (drawing).

4012. Horizontal Plane
Present location unknown

A stump of iron gnomon remains in position.

Limestone - The fragment is approximately half a dial with seven hour lines (5,6,7,8,9,10,11) and three day curves preserved. The meridian extends from the top of the fragment to some distance beyond the winter solstice.

Found in 1961 by Mr. Charles Leva at the Gallo-Roman village of Evelette near Huy in Belgium.

H = 275 mm
D = 20 mm

gs = 7 mm
se = 9 mm
ew = 32 mm

ge_7 = 17 mm
ge_8 = 21.5 mm
ge_9 = 28.5 mm
ge_{10} = 45 mm
ge_{11} = 91 mm

For $\varepsilon = 24°$, $\phi = 15°30'$
h = 12 mm
fs = 6 mm
A_3 = 64°
Q_3 = 25 mm

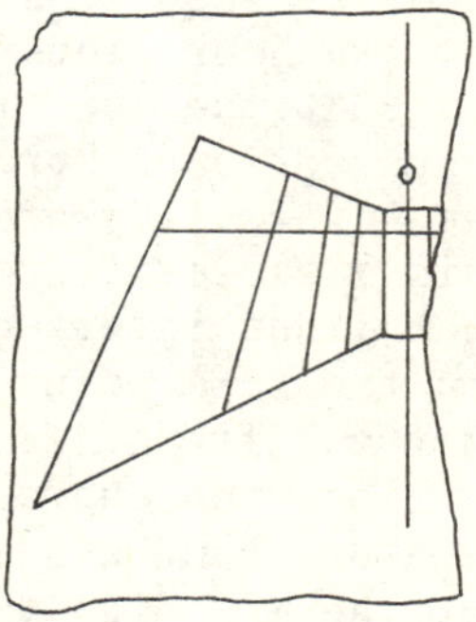

Michel, Henri. "Un Cadran Solaire Gallo-Romain." *Ciel et Terre* 82 (1966): 3-7, figure 1 (drawing).

4013. Horizontal Plane
Liége, Belgium
Musée Curtius

H = 320 mm
W = 250 mm
D = 140 mm

se = 21 mm
ew = 67.5 mm

Freson reports seeing hole along meridian 11 mm from summer solstice suitable for holding a perpendicular gnomon. This hole is not visible in the published photograph.

Freestone - Parts of eight hour lines are preserved on this triangular fragment (3,4,5,6,7,8,9,10). Complete hour lines do not extend beyond solstices, but meridian does continue beyond these bounds. Solstice curves seem to be approximate; equinox and hour lines are creditably straight. At the junction of meridian and engraved lower border of the fragment an interesting group of engraved lines is visible. It consists of two straight lines intersecting the meridian in the same point at angles of 24° on the left and 22° on the right. In addition, two engraved circles with centers can be seen on either side of the meridian tangent to these straight lines and the meridian.

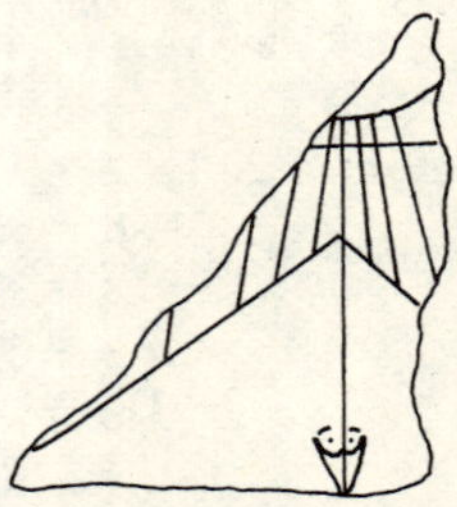

Found November 4, 1931, at Awans-Bierset near Liége.

For $\varepsilon = 24°$, $\phi = 50°$
h = 30 mm
fs = 14 mm

Freson, Abbe E. "Un Cadran Solaire Gallo-Romain." *Comptes Rendues du Congrès de la Fédération Archéologique et Historique de Belgique, 29 session*, Liége, 1932, number 4, pp. 124-29, plate 11 (photograph).

4014G. Horizontal Plane
Wiesbaden, Germany
Städtisches Museum, Sammlung Nassauischer Altertumer, no. 386

The round gnomon hole, 8 mm in diameter, contains traces of bronze and iron.

Limestone - A hole for possible mounting is visible in the back at top center. The stone is shaped like a truncated pyramid with the dial engraved on the base with the larger surface area. Eleven hour lines and three day curves are visible. Lines for the fifth and seventh hours split into two lines just below the equinox.

Found in 1867 in Wiesbaden in the ruins of a Roman bath.

H = 530 mm
W = 480 mm
D = 110 mm

gs = 16 mm
se = 23 mm
ew = 75 mm

ge_1 = 187 mm
ge_2 = 95 mm
ge_3 = 61 mm
ge_4 = 48 mm
ge_5 = 40 mm
ge_6 = 39 mm

$\kappa = 79°$
$\mu = 66°$
$\nu = 52°$
$o = 37°$
$\pi = 19°$
$\pi' = 11°$

For $\varepsilon = 24°$, $\phi = 50°$
h = 33 mm
fs = 16 mm
$A_3 = 52°$
Q_3 = 64 mm

Diels, p. 182, plate 15 (photograph).
Schalk, H., "Die neuesten Funde in Wiesbaden," AVN 9 (1868): 358 (drawing).
Schlieben, "Romische Sonnenuhren in Wiesbaden und Cannstadt," AVN 20(1888): 316-33 (drawings).

4015. Horizontal Plane
Mons Claudianus, Egypt
Lost

Listed as quarry mark from Mons Claudianus in the eastern desert of Egypt.

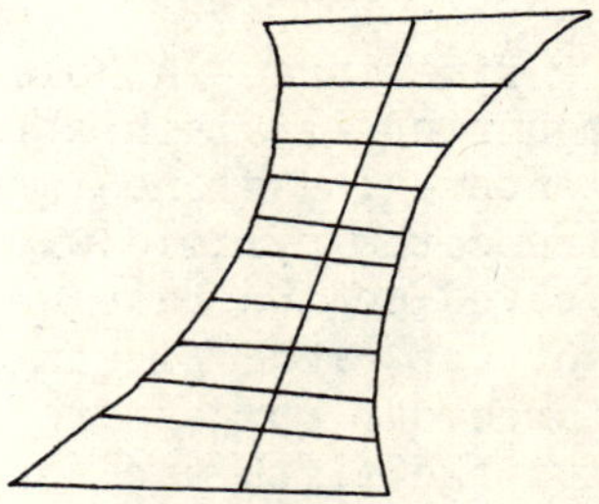

Meredith, D., *Chronique d'Egypte* 29 (1954): 119, no. 49 (drawing).
[corrected entry] Meredith, D., *Chronique d'Egypte* 30 (1955): 129 (drawing).

Meridian Dials

Measured Quantities

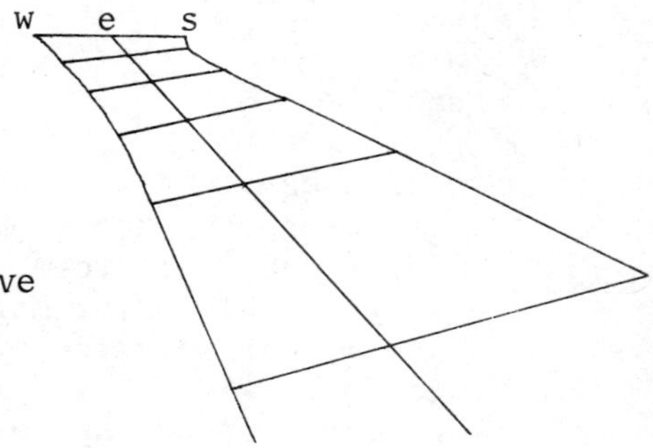

H = height
W = width
D = depth
d = diameter of disk

we = distance along horizon line from the winter solstice curve to the equinox
es = distance along the horizon line from the equinox to the summer solstice curve

(0 to III)
e = distance along the equinox from the horizon line to the third hour point

β = angle of deviation
α = angle between the horizon line and the equinox

Calculated Quantities

ϕ = intended latitude

h = perpendicular height of gnomon

C_3 = hectemoros arc for the third hour at the equinox

Q_3 = shadow length for the third hour at the equinox measured from the center of the gnomon hole

Vertical Dials

Measured Quantities

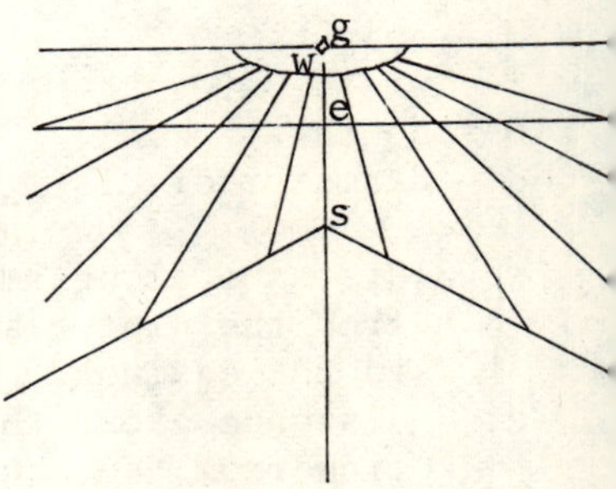

H = height
W = width
D = depth

gw = distance along the meridian from the center of the gnomon hole to the winter solstice

we = distance along the meridian from the winter solstice curve to the equinox

es = distance along the meridian from the equinox to the summer solstice curve

(III to VI)
e = distance along the equinox from the third to the sixth hour points

β = angle of deviation
μ = angle between the sixth hour line and the equinox
α = angle between the top edge and the equinox

Calculated Quantities

ε = obliquity of ecliptic
ϕ = intended latitude

h = perpendicular length of gnomon
fw = shortest noon shadow measured from the projection of the gnomon point on the dial face

P_3 = horarius arc for the third hour at the equinox

V_3 = vertical arc for the third hour at the equinox

Q_3 = shadow length for the third hour at the equinox measured from the center of the gnomon hole

Deviating Dials

Measured Quantities

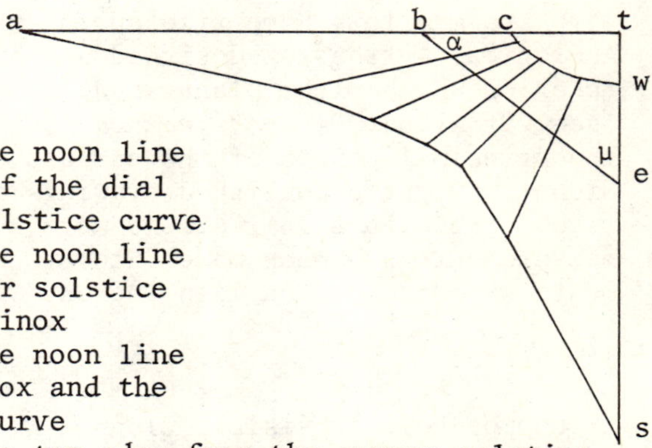

H = height
W = width
D = depth

tw = distance along the noon line between the top of the dial and the winter solstice curve
we = distance along the noon line between the winter solstice curve and the equinox
es = distance along the noon line between the equinox and the summer solstice curve
ab = distance along the top edge from the summer solstice to the equinox
bc = distance along the top edge from the equinox to the winter solstice

(VI to VII)
e = distance along the equinox from the sixth to the seventh hour lines

η = angle between the left and right faces of a double deviating dial
μ = angle between the sixth hour line and the equinox
α = angle between the top edge and the equinox

Calculated Quantities

ϕ = intended latitude
β = angle of deviation from prime vertical plane
ξ = ortive amplitude
$\overline{\psi}$ = angle between the axis of the earth and its projection on the deviating plane of the dial face

h = perpendicular length of gnomon
go = oblique length of gnomon

5001. Vertical Planes
Athens, Greece
Roman Agora

White marble - Some time during the first century B.C. Andronicus Cyrrhestes designed and built an octagonal tower on which eight planar sundials are visible today (see Vitruvius, *De architectura* I, 6.4). The dials face the principal points of the compass. James Stuart and Nicholas Revette published drawings and measurements of five of the dials in 1762. No more recent attempts at measurement have been made. Stuart and Revette's measurements were made in English inches.

South Face

Gnomon hole is visible along meridian above horizon line.

H = 88.05"
W = 64.034"

Eleven hour lines and three day curves are preserved. An allegorical relief and the Greek inscription νότος are visible above the dial face.

gw = 6.025"
we = 17.825"
es = 29.7"

(III to VI)
e = 17.634"

For $\varepsilon = 24°$, $\phi = 37°20'$, h = 10.25"
fw = 5.6" and
$(\text{III to VI})_e$ = 16.88"

$\beta = 0°$

East Face

Five hour lines mark the first to the fifth hours of the day between the engraved summer and winter solstices. An engraved equinox crosses the hour lines between the solstices. An allegorical relief and the Greek inscription ἀπηλιώτης decorate the dial face.

H = 143.55"
W = 126.284"

we = 10.9"
es = 11.1"

$(\text{0 to III})_e$ = 19.20"
$(\text{0 to I})_e$ = 5.05"
$(\text{I to II})_e$ = 6.3"

$\phi = 38.8°$
h = 19.20"

$\beta = 90°$
$\alpha = 51.2°$

5001. continued

Southeast Face

A square gnomon hole is visible along the horizon at a point 9.85" from the meridian.

Five hour lines are engraved. The meridian coincides with an edge of the octagonal tower. Solstices and equinox cross the hour lines. An allegorical relief and the Greek inscription εὖρος are visible above the dial face.

For $\varepsilon = 24°$, $\phi = 38°11'$, $\beta = 46°6'$
$h = 25.35''$ and
$(V \text{ to } VI)_e = 21.37''$.

H = 265.23"
W = 128.05"

tw = 18.834"
we = 27.2"
es = 98.12"

$(III \text{ to } IV)_e = 9.2''$
$(IV \text{ to } V)_e = 12.166''$
$(V \text{ to } VI)_e = 20.225''$

$\beta = 45°$
$\mu = 47.5°$

Northeast Face

The gnomon hole is positioned along a line perpendicular to the horizon line which intersects the equinox.

Four engraved hour lines and a horizon line are crossed by the winter solstice, the equinox, and the summer solstice. One hour line intersects the winter solstice; three hour lines intersect the equinox; and four hour lines intersect the summer solstice. An allegorical relief and a Greek inscription καικίας decorate the dial face.

H = 125.95"
W = 127.10"

$se_{12} = 5.45''$

$ew_{12} = 18.7''$

$\beta = 225°$
$\alpha = 41°$

5001. continued

North Face

A gnomon hole can be seen along the meridian line above the horizon.

H = 33.6"
W = 129.50"

β = 360°

Two hour lines and the summer solstice are engraved. An allegorical relief and a Greek inscription βορέας decorate the block above the shadow-receiving surface.

Northwest Face

Stuart and Revette took no measurements of this face, but the dial is similar to that engraved on the Northeast Face. An allegorical relief and the Greek inscription σκίρων decorate the block above the shadow-receiving surface.

West Face

Stuart and Revette took no measurements of this face, but the dial is similar to that engraved on the East Face. An allegorical relief above the shadow-receiving surface is inscribed [ζ]έ[φυρος].

Southwest Face

Stuart and Revette took no measurements of this face, but the dial is similar to that engraved on the Southeast Face. An allegorical relief above the shadow-receiving surface in inscribed λίψ.

For other sundials inscribed with wind names see 4002G, 4008G, 4009, and 4010.

5001. continued

Convex Cylinder (Below South Face)

A small partially preserved cylindrical chamber with radius equal to 148 cm extends from the south face of the octagonal tower. On the outer surface of the cylinder, two sets of intersecting lines (meeting at angles of 52°) are visible engraved at the sides, perhaps forming part of a ninth dial. In the figures below ABC is the horizontal section of the cylinder. DD" is the lateral surface of the cylinder showing the engraved lines.

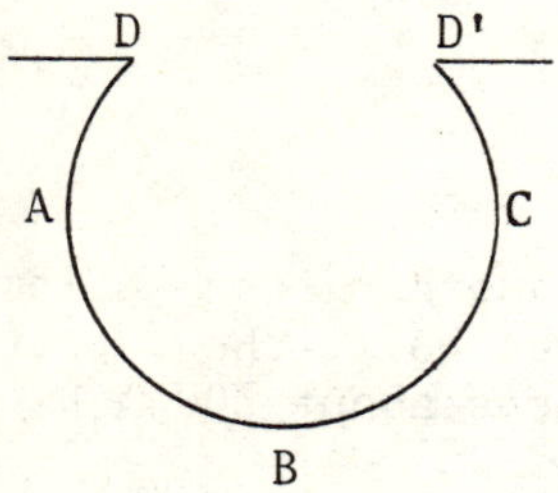

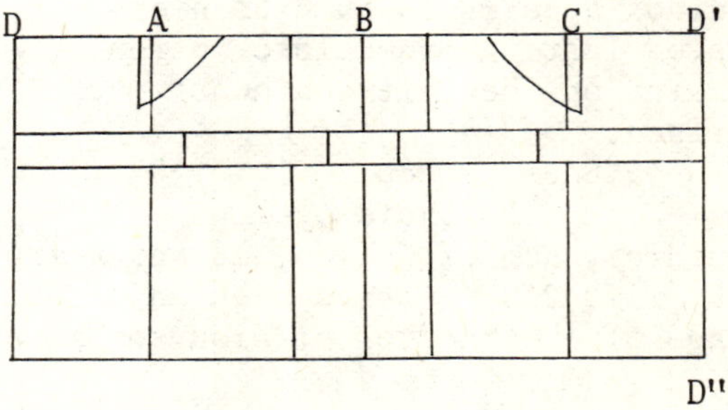

Delambre, J. B. J., *Histoire de l'Astronomie Ancienne*, Paris, 1817, pp. 487-504 and plates 11 and 12 (drawings).
Stuart, J., and Revette, N., *The Antiquities of Athens*, vol. 1, London, 1762, pp. 13-25 (drawings with measurements).
Cf. DA, vol. 3, p. 259, figure 3887 (drawing).
Noble, J. V., and Price, D. J., "The Water-Clock in the Tower of the Winds," AJA 72(1968): 345-55, plates 111-18 with photographs.
Robinson, H. S., "The Tower of the Winds and the Roman Market-Place," AJA 47(1943): 291-305.

5002G. Vertical Plane
London, England
British Museum, no. 2544

The position and measurements of the gnomon holes are shown in the accompanying diagram. No metal remains.

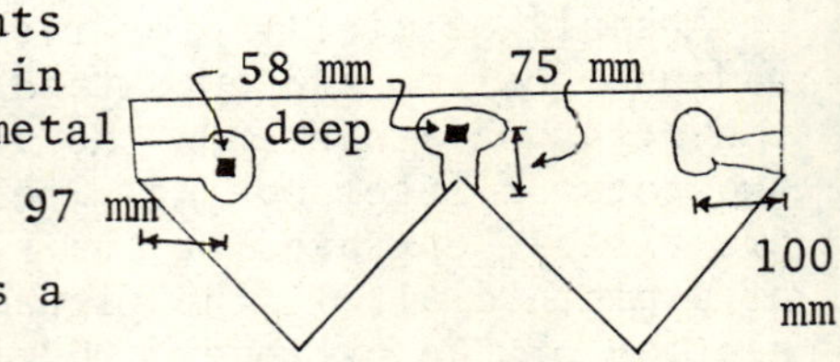

Cream marble - This dial is a combination of four deviating vertical dials. On all faces, the hour lines have been engraved less heavily than the day curves. The equinox appears to have been engraved last. A Greek inscription remains on the left southwest-facing surface:

Φαῖδρος Ζωΐλου
Παιανιεὺς ἐποίει.

H = 497 mm
W = 985 mm
D_{max} = 323 mm

Phaidros, son of Zoilos, is known from another inscription (IG, III, 239) to have been an archon who built the latest stage of the theater of Dionysos at Athens about 300 A.D.

Left Southwest Face

Five engraved hour lines, not including the meridian, and three day curves have been preserved.

Delambre computes h = 113.4 mm, for $\phi = 37°30'$, $\varepsilon = 23°51'$, and we = 118.9 mm.
For $\beta = 45°$ and $\mu = 46°$, $\phi = 36°13'$.

H = 497 mm
W = 343 mm

tw = 77 mm
we = 119 mm

(VI to VII)
e = 96 mm
(VII to VIII)
e = 52 mm

$\mu = 46°$

5002G. continued

Left Southeast Face

Five engraved hour lines, not including the meridian, and three day curves are preserved. The meridian is formed by the intersection of left southeast face and right southwest face.

Delambre computes h = 113.6 mm, for $\phi = 37°30'$, $\varepsilon = 23°51'$, and we = 119.2 mm.
For $\beta = 45°$ and $\mu = 49°$, $\phi = 39°8'$.

H = 497 mm
W = 351 mm

tw = 79 mm
we = 119 mm

(VI to VII)
e = 87 mm
(VII to VIII)
e = 59.5 mm

$\mu = 49°$

Right Southwest Face

Four engraved hour lines, not including the meridian, and three day curves are preserved. The meridian is formed by the intersection of the left southeast and the right southwest face.

For $\beta = 45°$ and $\mu = 47°$, $\phi = 37°10'$.

H = 497 mm
W = 351 mm
tw = 79 mm
we = 119 mm

(VI to VII)
e = 90 mm
(VII to VIII)
e = 56 mm

$\mu = 47°$

Right Southeast Face

Five engraved hour lines, not including the meridian, and three day curves are preserved. The meridian is formed by the right edge of the block.

For $\beta = 45°$, and $\mu = 48°$, $\phi = 38°9'$.

H = 497 mm
W = 343 mm
tw = 79 mm
we = 108 mm
(VI to VII)
e = 92 mm
(VII to VIII)
e = 55.5 mm
$\mu = 48°$

Delambre, J. B. J., *Histoire de l'Astronomie Ancienne*, vol. 2, Paris, 1817, pp. 504-09.
IG, III, 427.
Cf. Gatty, pp. 40-41 (drawing).
Smith, pp. 412-13.

5003G. Vertical Plane
Athens, Greece
Museum of the Stoa of Attalos, study collection, no. A1869.

H = 150 mm
W = 500 mm
D = 215 mm

Hymettian marble - Eleven scratched hour lines are connected by two semicircles at the top and bottom; hour lines do not extend beyond curved boundaries. The left edge is finished smoothly at the top and bears a trace of curved molding. With the help of a strong light it is possible to see scratched Greek letters near the outer circle:

Α Β Γ Δ Ε

The angles formed by the hour lines are from left to right: 17.5°, 15.5°, 15°, 13°, 14.5°, 15.5°, 14.5°, 14°, 15.5°, 11.5°, 15.5°, and 18°.

Found June 25, 1951, in a marble pile north of the Temple of Ares in the Greek Agora, Athens.

Unpublished

5004G. Vertical Plane
Ancient Corinth, Greece
Odeum, South Wall

Porous stone - Seven hour lines, meeting at a hole at the top, and joined by a broken line resembling a quarter circle at the bottom, form this dial grid. A scraped area to the right of the meridian may indicate either preparation for the completion of the dial or destruction of part of the grid. Broneer remarks that the ruined odeum at Corinth was used for private dwellings in the third and fourth centuries A.D.; the scratched dial probably dates from this period. The hour lines measure from left to right: 125 mm, 119 mm, 118 mm, 122 mm, 119 mm, 125 mm, 130 mm. These hour lines form the following angles, measured from left to right: 15.5°, 14.5°, 13°, 10.5°, 14°, 15.5°.

Block:
H = 270 mm
W = 800 mm

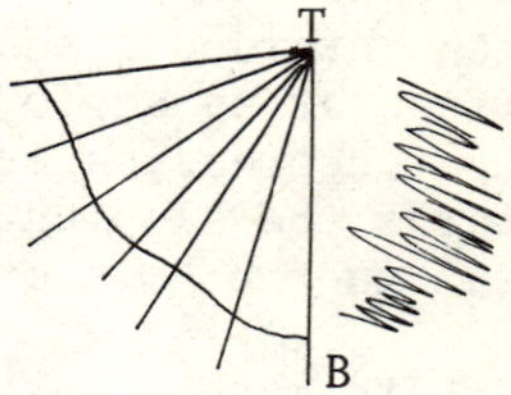

line TB is plumb

Broneer, O., *The Odeum*, Cambridge: Harvard University Press, 1932, p. 141, figure 137 (drawing).

5005G. Vertical Plane
Oropus-Amphiareion, Greece
Archeological Site, museum portico

Plate 56
Page 352

$H = 822$ mm
$W_{max} = 363$ mm
$D = 75$ mm
$tw = 88$ mm
$we = 127$ mm
$es = 468$ mm
(VI to VII)
$e = 95$ mm
(VII to VIII)
$e = 55$ mm
$\mu = 45°$

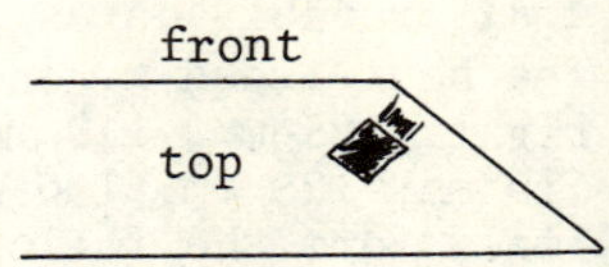

A gnomon hole 18 mm square and 18 mm deep is preserved in the top surface of the dial above the meridian. Its sides are oriented parallel to and perpendicular to the sloping edge which intersects the edge of the dial surface at an angle of 142.8°. A slight depression leads from this hole towards the sloping edge. There is no metal in the hole or the groove.

White marble - Three fragments have been joined to form this southwest facing dial surface. The finished left edge of the slab is formed by a sloping surface which intersects the dial surface at an angle of 142.8°. Top and bottom edges are smoothly finished and a section of the right edge smoothly follows the curve of the summer solstice. Five engraved hour lines have been preserved in addition to the meridian which is formed by the intersecting planes of the left edge. Three day curves extend from the meridian to the top edge. Two clamp holes and grooves are preserved in the rough picked back along the unbroken meridian edge.

For $\varepsilon = 24°$, $\phi = 37°50'$
$\beta = 50°57'$
$h = 106.38$ mm
VI to VII_e computed = 102 mm

Two other vertical dials found at Oropus-Amphiareion are 5006G and 5007G.

Unpublished

5006G. Vertical Plane
Oropus-Amphiareion, Greece
Archeological Site, museum portico

Plate 57
Page 352

White marble - This fragment is the upper left corner of a southeast-facing dial. Top and lower edges are finished smooth. Three engraved hour lines (1,2,3) and three day curves are preserved. All planes of the slab surface intersect at right angles.

PH = 422 mm
PW = 166 mm
D = 63 mm

bc = 80 mm
ab = 266 mm

(0 to I)
e = 42mm
(I to II)
e = 35 mm

$\alpha = 44°$

For $\varepsilon = 24°$, $\phi = 37°36'$

$\beta = 48°2'$
$\xi = 30°53'$
h = 113.6 mm
$(o\ to\ I)_e$ computed = 40.43 mm

Unpublished

***

5007G. Plane Fragment
Oropus-Amphiareion, Greece
Archeological Site, museum portico

Cream and pink stone - lines lightly engraved.

H = 130 mm
W = 166 mm
D = 64.5 mm

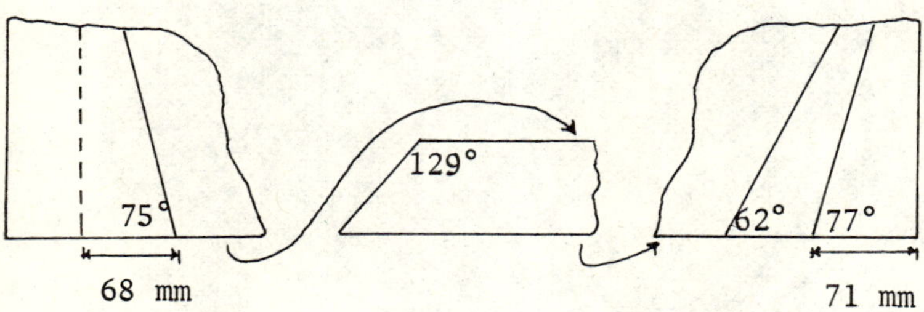

Unpublished

55

56

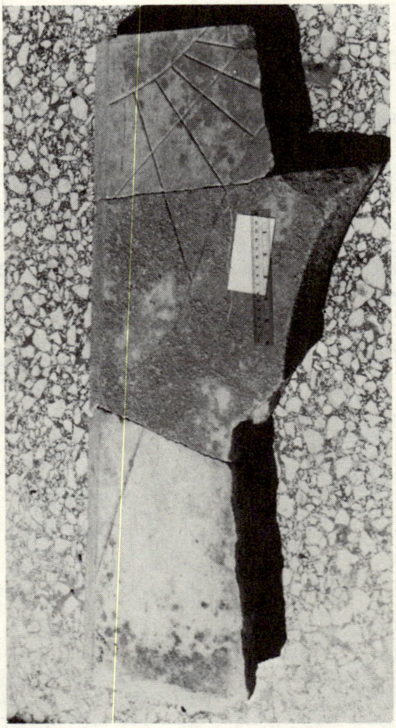

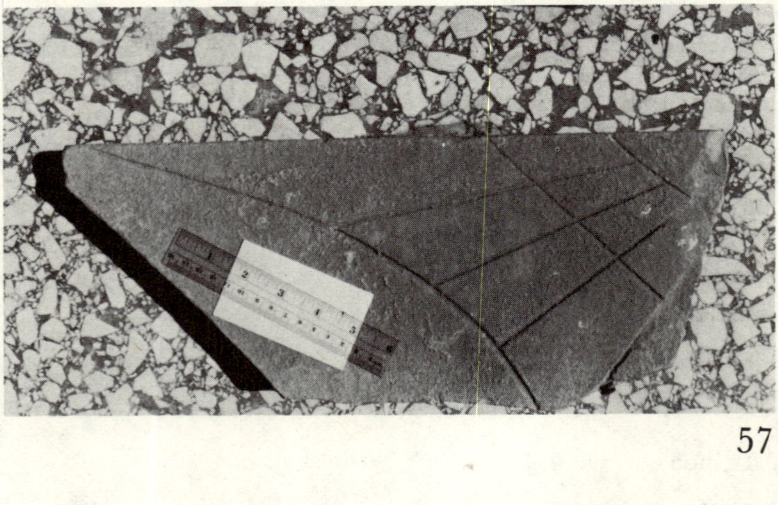

57

5008G. Vertical Plane
Delos, Greece
Archeological Museum, study collection in portico

The base of a bronze gnomon in lead fill remains in the top surface above the meridian formed by the intersecting dial surfaces.

White marble - This well-preserved dial consists of two dial surfaces, each with five engraved hour lines and three day curves, meeting at the meridian; one surface faces southeast, the other southwest. Some hour lines on the right extend beyond the summer solstice, but all seem to be contained within the winter curve.

Found south of the oikos of Dionysos on Delos.

H = 295.5 mm
W = 210 mm
D = 140 mm

tw = 26.5 mm
we = 33 mm
es = 127.5+ mm

(VI to VII)
e = 32.5 mm
(VII to VIII)
e = 16.5 mm

$\eta = 80°$
$\alpha = 45.2°$

Deonna observes (p. 195) that the engraved lines of the equinox and the summer solstice continue onto the top cross section. I did not observe these lines.

For $\beta = 50°$ and $\varepsilon = 24°$,
$\phi = 37°15'$
h = 27.7 mm
go = 43 mm
$(\text{VI to VII})_e$ computed = 27.6 mm

Deonna, pp. 194-95, figs. 232-33 (drawings), plate 545 (photograph).

5009. Vertical Plane
Delos, Greece
Lost

White marble - The drawing in Deonna shows fourteen hour lines. Part of an engraved circular curve is visible at lower left. The hour lines radiate from the gnomon hole to the circular arc.

Found in a house southeast of the lake on Delos.

H = 140 mm
W = 160 mm

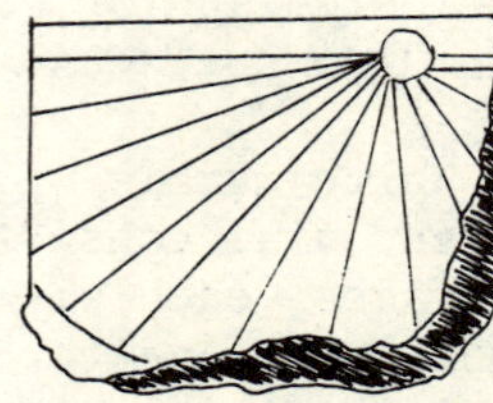

Deonna, p. 194, fig. 231 (drawing).

***

5010. Vertical Plane
Present location unknown

Marble - Three day curves and six hour lines (including horizon line) are visible. Found on the island of Delos near the Temple of Apollo. Reported to be in the "cabinet des antiques," Paris, in 1814.

d = 20 English inches

$\alpha = 59.5°$

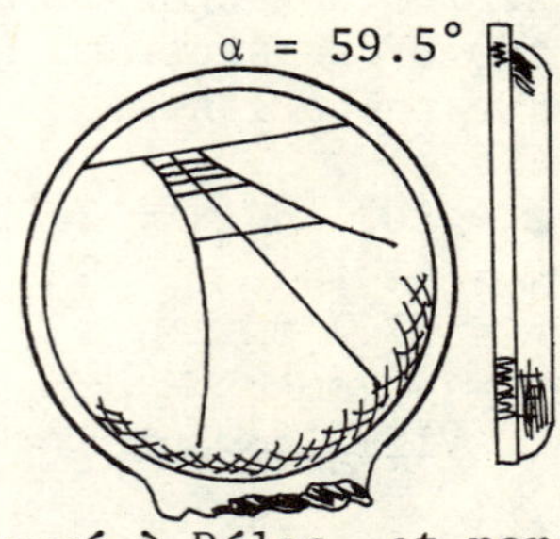

Delambre, J. B. J., "D'un Cadran Trouvé à Délos, et par Occasion de la Gnomonique des Anciens," *Analyse des Travaux de la Classe des Sciences Mathematiques et Physiques de l'Institut Royal de France*, Paris, 1814, pp. 29-39.
Kockerell, C. K., letter and drawing dated August 28, 1815, published with Francesco Peter's *Di un Antico Orologio Solare*, Rome, 1815.

5011G. Vertical Plane
Altino, Italy
Museo Archeologico

A hole for a gnomon is preserved in the top of the dial directly above the meridian line; some residue of lead remains.

White stone of Histria - A smooth curved edge on the right side follows the curve of the summer solstice. Five fairly deeply incised hour lines do not continue beyond either solstice curve. The right edge is finished in a double concave molding. There is no trace of color, relief decoration, or inscription.

Found October 15, 1969, in the necropolis of Altino. It is probably part of one of the cemetery monuments dating from the beginning of the first century A.D.

For ε = 24°, ϕ = 44°6'
β = 31°45'
h = 175.5 mm
$(\text{VI to VII})_e$ computed = 85.9 mm

H = 540 mm
PW = 360 mm
D = 75 mm

tw = 56 mm
we = 130 mm
es = 351+ mm

(VI to VII)
e = 92 mm
(VII to VIII)
e = 57 mm

μ = 61.5°

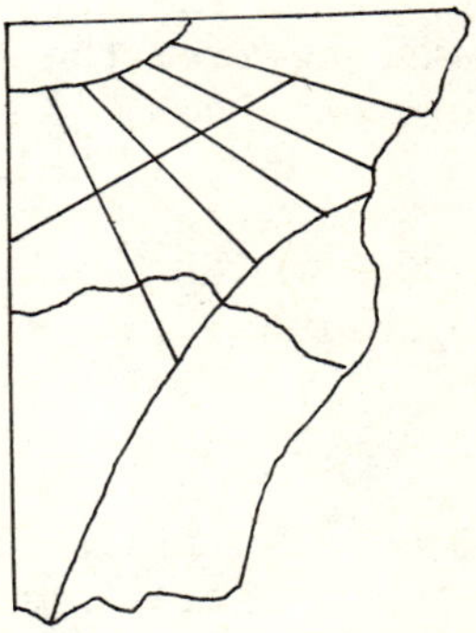

Unpublished

5012G. Vertical Plane
Ostia Antica, Italy
Piccolo Mercato, Lapidario

Plate 58
Page 364

The combination of grooves at the top center of this dial block presumably held the gnomon.

H = 143 mm
W = 324 mm
D = 27 mm

d = 264 mm

Cream crystalline marble - Eleven fairly deeply scratched hour lines divide a single semicircle into twelve parts on this dial. The shadow-receiving surface is uniformly thick and smoothly finished. The hour lines measure from left to right: 130 mm, 131 mm, 132 mm, 132 mm, 132 mm, 130 mm, 134 mm, 135 mm, 136 mm, 135 mm, 132 mm. They form the following angles, measured from the left to the right horizon: 10.8°, 18°, 15°, 15.4°, 15.8°, 15°, 12°, 15.8°, 16°, 20.5°, 16°, 9.7°.

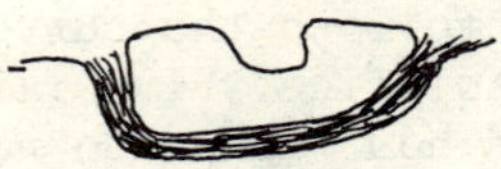
Front

Top

Back

Unpublished

***

5013G. Vertical Plane
Ostia Antica, Italy
Piccolo Mercato, Lapidario

White Greek marble - Five deeply engraved hour lines, the equinox, and the winter solstice remain on this dial fragment. The meridian is not preserved.

H = 233 mm
W = 135 mm
D = 45 mm

(VII to VIII)
e = 53 mm

Unpublished

5014G. Vertical Plane
Ostia Antica, Italy
Magazine, no. 1892

H = 420 mm
W = 256 mm
D = 64 mm

tw = 85 mm
we = 123 mm

(IV to V)
e = 60 mm
(V to VI)
e = 93 mm

$\mu = 58°$

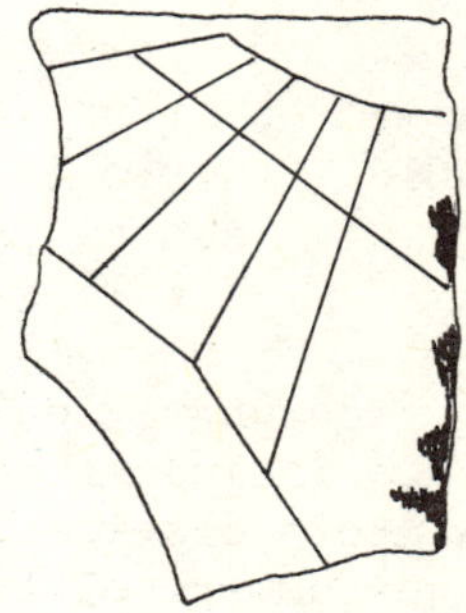

Two preserved holes may have served separately or together as gnomon holders. One hole on the top of the dial is filled with lead and connected to the right edge by a groove 85 mm long, 23 mm wide and 3.5 mm deep. A second hole, 10 mm wide, is situated 34 mm from the top along the right edge sloping up at an angle of 45°.

Colonial marble, cream - Six engraved hour lines, including the meridian, are preserved on this southeast-facing half of a vertical deviating dial. The equinox and two solstices do not extend beyond the first hour line. Although the dial has been squared off along the left edge, a portion of the curved border remains. The top and right edge are smoothly perpendicular to the dial face. A large engraved *B* is preserved at the left end of the top edge. Evidence at the junctures of the wide hour lines and the equinox indicates that the former were engraved last.

Found May 21, 1962, being used as a paving stone along the Decumanus near the baths.

For $\varepsilon = 24°$, $\phi = 40°10'$
$\beta = 36°14'$
$h = 141.62$ mm
$(\text{VI to VII})_e$ computed = 84.6 mm

Vertical dial 5013G was also found at Ostia.

Unpublished

5015. Vertical Plane
Pompeii, Italy

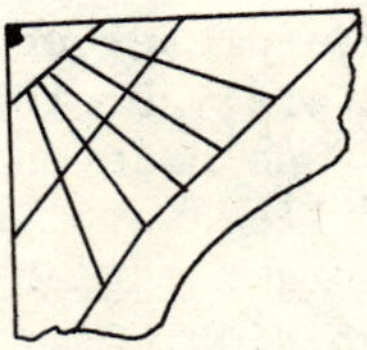

A fragment of vertical deviating sundial is visible in a photograph of "Sundials at Pompeii" (third from left) in the *Enciclopedia Italiana*. The dial could not be located for examination in July, 1971.

Gabba, L. and Jacono, L., "Meridiana," *Enciclopedia Italiana*, Rome, 1935-43, p. 303.

***

5016. Vertical Plane
Seen at Casa Deangelis,
Tivoli, Italy in 1798

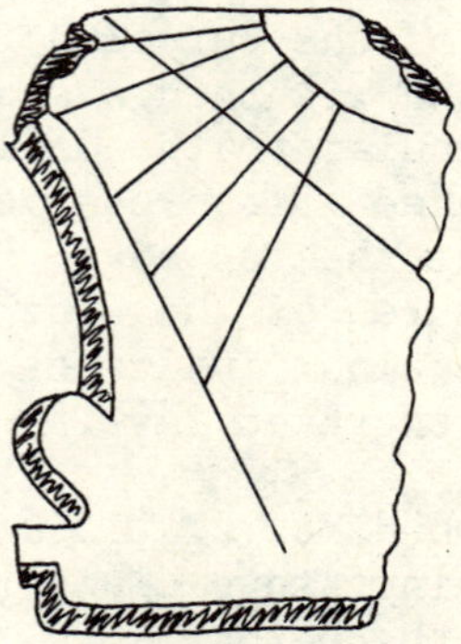

The engraved lines on this dial face include five hour lines, the winter solstice, the equinox, and the summer solstice. The dial appears to have been made to face southeast.

Antonini, no. 10.

5017. Vertical Plane
Palestrina, Italy
South front of the Cathedral of St. Agapito

Marucchi describes a configuration of four engraved lines which he suggests served to mark the positions of the third and ninth hours of the day between winter and summer solstice.

Marucchi, O., "Nuovi Studi Sull'Antichissimo Orologio Solare di Palestrina," *Rendiconti della Pont. Accad. Rom. di Archaeol.* 6(1928): 77-84 and plate 1.
Cf. *Archäologischer Anzeiger* 44 (1929): 119, col. 1.

***

5018G. Vertical Plane
Syracuse, Italy
National Archeological Museum, no. 45870

A notably large gnomon hole, 13 mm wide and 45 mm deep, is preserved in the top center of the stone.

H = 145 mm
W = 172 mm
D = 30 mm

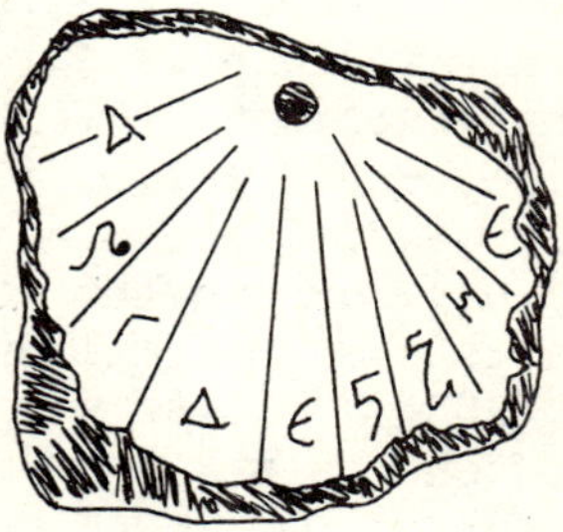

White stone - Greek letters mark the nine divisions determined by ten scratched hour lines on this dial.

Palazzolo, Italy, is the supposed provenance of the fragment.

Unpublished

5019G. Vertical Plane
Brugg, Switzerland
Vindonissa Museum, nos. 7296, 7297, 7298

D = 20 mm

White limestone - A blood red color marks the eleven preserved hour lines and three day curves of this dial. There is no red color along the edges near the missing meridian line, and in fragment 7297 the left or "meridian" edge is finished smooth. The smooth lower edges of fragment 7297 also give some indication of the original shape of the dial. All preserved hour lines continue in red to the top center of the dial where a groove remains in fragment 7296. The two engraved hour lines in fragment 7297 continue beyond the lowest day curve without red color. Note that while the hour lines are straight lines, the day curves are broken lines, the top curve most closely resembling a semicircle. The hour lines form the following angles between the left horizon and the eleventh hour: 19°,18°,15.5°, 17.2°,14.1°,11°,10°,11°,14.7°,12.8°, 17.5°. The winter solstice curve consists of the following arcs: 9, 11, 9, 10, 8,...,..., 8, 11, 8, 10,... mm. The equinox curve between the horizon and the third hour line consists of the following arcs: 25, 26, 20 mm. The summer solstice curve between the sixth and eighth hour lines measures: 37, 31 mm.

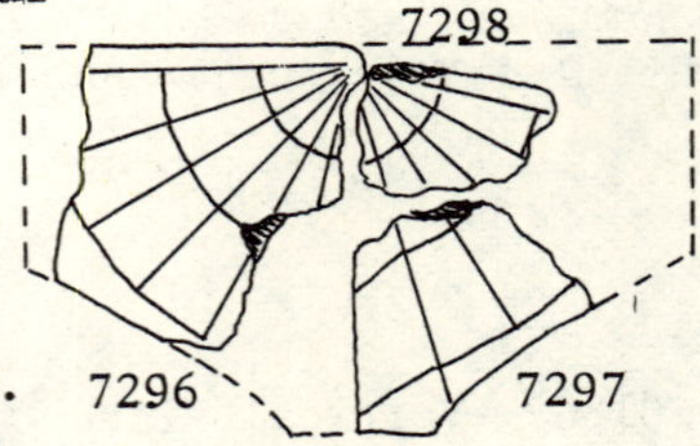

Simonett, C., *Führer durch das Vindonissa Museum*, Brugg, 1947, pp. 67-68 (drawing).

5020. Vertical Plane
Narbonne, France
Musée Lapidaire, Ancienne Église de Lamourguier, no. 869-I09-I

The publication describes a red copper bayonetlike gnomon.

White stone from Estaillades - The dial consists of two plane surfaces meeting at an angle of 90° to form the meridian. Eleven hour lines and three day curves are preserved.

The dial was found in 1869 during excavations of the Narbonne cemetery.

H = 774 mm
W = 340 mm
D = 64 mm

tw = 60 mm
we = 90 mm
es = 315 mm

(VI to VII)
w = 42 mm
e = 90 mm
s = 252 mm

$\eta = 90°$

For $\varepsilon = 24°$ and $\beta = 45°$,
$\phi = 38°43'$
$h = 86.9$ mm
$\mu = 48°35'$
$(\text{VI to VII})_e$ computed = 69.6 mm

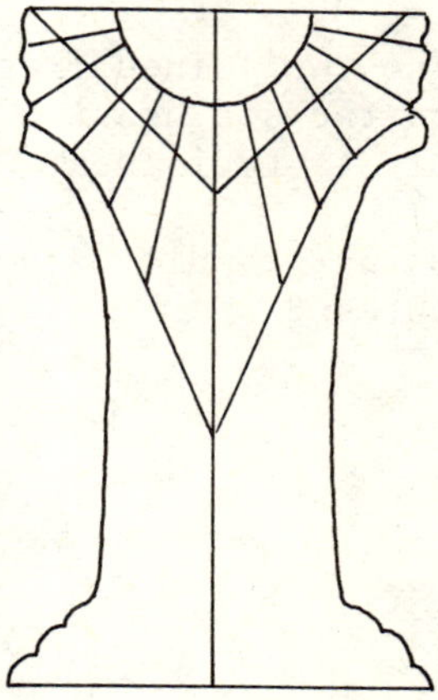

Bulletin de la Société Nationale des Antiquaires de France, Paris, 1896, pp. 225, 226 (drawing)
Measurements from the records of the Musée Lapidaire.

5021. Vertical Plane
Carthage, Tunisia
Musée Lavigerie de Saint-Louis, no. 1899

White marble - Five hour lines are preserved, not including the meridian. Parts of seven day curves remain. A Latin inscription marks these day curves as shown in the figure. The inscription reads:

H = 540 mm
W = 610 mm
μ = 45°

Brumalis
Januarius
Februarius
Marti[us]
Aprilis Aequinoctialis
[M]aius
[Ju]nius
[Solst]itialis

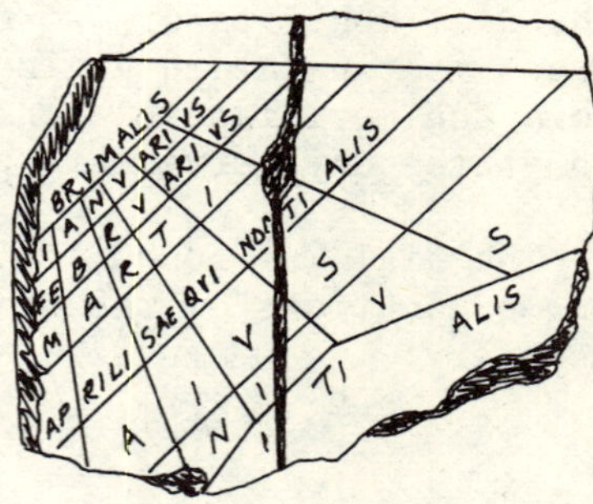

The two joined fragments were found on the southeast slope of the hill of St. Louis near Carthage.

Using graphic methods Tannery computes β = 48°.

CIL, VIII, 4, Supplement, 24677.
Delattre, R. P., *Musées de l'Algérie et de la Tunisie*, vol. 8, pt. 2: *Musée Lavigerie de Saint Louis*, Paris, 1899, p. 85 and plate 20, 5 (photograph).
Cf. Drecker, p. 54 and fig. 82 (photograph).
Tannery, P., *Mémoires Scientifiques*, vol. 3, Paris, 1915, pp. 1-11 (photograph).

5022G. Plane Surface — London, England — British Museum, no. 2546

Plate 59
Page 364

North Face

The preserved gnomon hole is 6 mm in diameter.

Parian marble - Three circular arcs (with the gnomon hole as the center) are each evenly divided by eleven hour lines. Arc c_1 is approximately 103°; arc c_2 is approximately 100°; arc c_3 is approximately 96°.

For $\varepsilon = 24°$, $h = 30.27$ mm and
c_1 corresponds to $\delta = 24°$
c_2 corresponds to $\delta = 21°5'$
c_3 corresponds to $\delta = 14°6'$
The dial grid appears to have been designed to have functioned in the plane of the equator, inclined at an angle of 32° from vertical.

H = 347 mm
W = 292 mm

tg = 16.5 mm
gc_1 = 70.5 mm
c_1c_2 = 10.5 mm
c_2c_3 = 42 mm
c_3b = 207.5 mm

(V to VI)
c_1 = 21 mm
c_2 = 23 mm
c_3 = 34 mm

South Face

The preserved gnomon hole is 6 mm in diameter.

Parian marble - A horizontal line crosses the dial surface 31.5 mm from the top edge. Three circular sections (with the gnomon hole as center) are evenly divided by eleven hour lines. Arc c_4 is approximately 75.8°; arc c_5 is approximately 77.6°; and arc c_6 is approximately 81°.

For $\varepsilon = 24°$, $h = 32.47$ mm and
c_4 corresponds to $\delta = -24°$
c_5 corresponds to $\delta = -21°22'$
c_6 corresponds to $\delta = -14°27'$.

H = 319 mm
W = 292 mm

tg = 16.5 mm
g to line = 15 mm
line to c_4 = 54 mm
c_4c_5 = 10 mm
c_5c_6 = 43 mm
c_6b = 180.5 mm

(V to VI)
c_4 = 15.5 mm
c_5 = 18 mm
c_6 = 30 mm

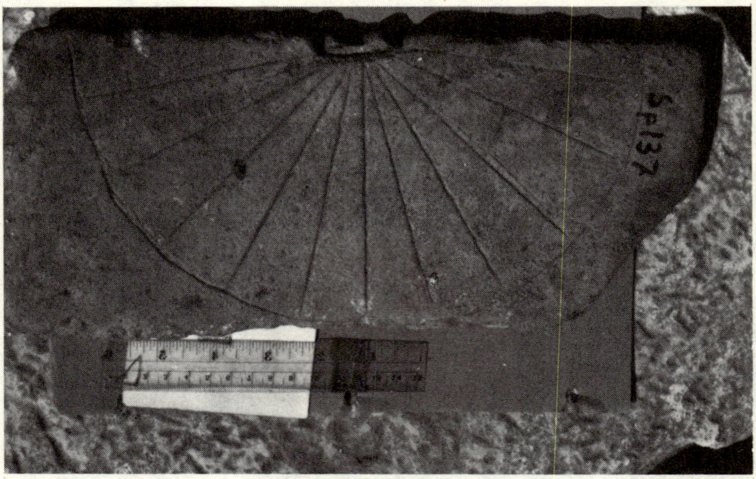

58

59

5022G. continued

West Face

A trace of metal remains in the gnomon hole (diameter is 4 mm).

H = 334.5 mm
W = 40 mm

A long straight line from the gnomon to the bottom of the slab is divided into six parts by five marks. The divisions measure from top to bottom: 12 mm, 16 mm, 19 mm, 33.5 mm, 87 mm, and 152 mm.

h = 47 mm

East Face

The empty gnomon hole has diameter 4.5 mm and is 16.5 mm deep.

H = 333.5 mm
W = 40 mm

A long straight line from the gnomon hole to the bottom of the slab is divided into six parts by five marks. The divisions measure from top to bottom: 12.5 mm, 16 mm, 20 mm, 33 mm, 87 mm, and 150 mm.

h = 48.5 mm

Smith, vol. 3, p. 414, no. 2546.

5023. Vertical Plane
Present Location Unknown

White stone - Seven hour lines radiate from a central point and cross four concentric circles. The preserved arcs of these circles are emphasized with red color.

H = 130 mm
W = 120 mm

Found in Samaria-Sebaste in 1908 near the altar of the Herodian temple.

Fisher, C. S., Lyon; D. S., Reisner, G. A., *Harvard Excavations at Samaria*, Cambridge, 1908-10, p. 26 and p. 15, no. 34.

***

5024. Vertical Plane
Rome, Italy
Museo Nazionale Romano, lost

Plates 60 (East) and 61 (West)

Photographs of this dial (negative numbers 54820 and 54821) indicate that it consists of a vertical disk facing east and west. Unfortunately attempts to locate it in the museum's study collection on the summer of 1971 were unsuccessful.

East Face:
$\alpha = 48°$

West Face:
$\alpha = 47°30'$

Unpublished

60

61

5025G. Vertical Plane
London, England
British Museum, no. E63

White marble - The unbroken lower left edge of this southeast-facing fragment follows the curve of the summer solstice. Parts of all three day curves and one hour line are broken away at the bottom and the right side. All or part of five hour lines remain. A group of five convex flutes decorates the top edge.

Provenance unknown.

PH = 454 mm
PW = 227 mm
D = 73 mm
bc = 75 mm
ab = 229 mm
(0 to I)
e = 41 mm
(I to II)
e = 34.5 mm
α = 42.6°

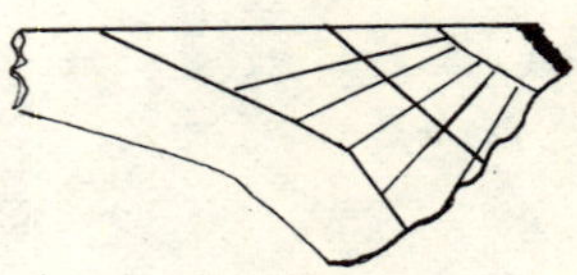

For ε = 24°, ϕ = 40°17'
β = 51°12'
ξ = 32°13'
h = 108.9 mm
$(0 \text{ to } I)_e$ computed = 37.03 mm

Compare 5006G, also the upper left corner of a deviating dial.

Unpublished

Cylindrical Dials

Measured Quantities

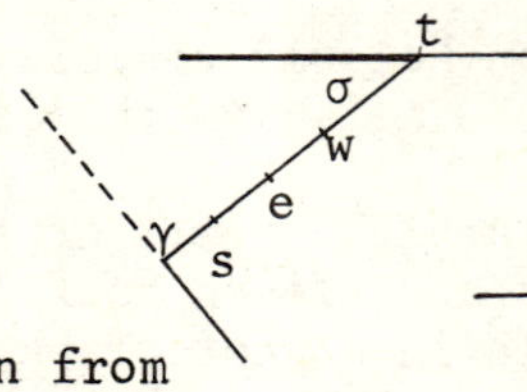

H = height
W = width
D = depth
d = diameter of cylinder

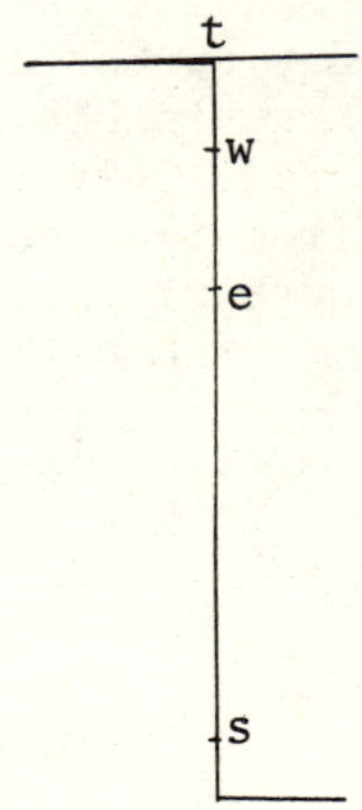

tw = distance along the meridian from the top edge of the cylinder to to winter solstice
we = distance along the meridian from the winter solstice to the equinox
es = distance along the meridian from the equinox to the summer solstice.
sb = distance along the meridian from the summer solstice to the lower edge of the cylinder

(VI to VII)
- w = distance along the winter solstice curve from the sixth to the seventh hour points
- e = distance along the equinox from the sixth to the seventh hour points
- s = distance along the summer solstice curve from the sixth to the seventh hour points

σ = angle between the dial face and the horizon, measured in the meridian plane
γ = angle between the dial face and the oblique front of the dial, measured in the meridian plane

Calculated Quantities

ϕ = intended latitude
ε = obliquity of ecliptic

h = gnomon length
oe = radius of the cylinder

6001. Cylindrical
Volubilis, Morocco
Archeological Museum

A gnomon hole is visible in the top of the dial above the meridian.

White marble - Parts of three day curves and eleven hour lines are visible in a photograph of the dial. Only the meridian continues from the gnomon hole to the bottom edge of the dial. Other hour lines stretch between the solstices.

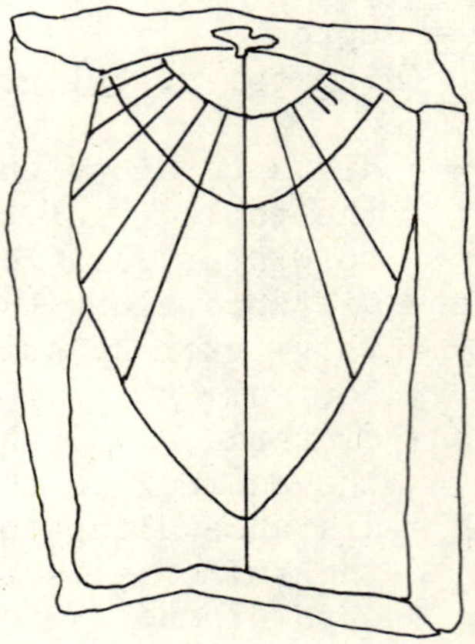

Drawing taken from a photograph of the dial taken by M. Henri Michel in 1952.

6002G. Cylindrical
Constanta, Rumania
Archeological Museum, no. 1657

Some lead remains in the gnomon hole in the top surface of this dial.

H = 494 mm
W = 300 mm
D = 160 mm

tw = 102.5 mm
we = 47 mm
eb = 49 mm

(VI to VII)
w = 28 mm
e = 40.2 mm
s = 52 mm

$\sigma = 73.8°$
$\gamma = 47°$

White marble - The shadow-receiving surface of this dial is supported by a bull's head and horns. The rounded back of the dial has the shape of a "feminine idol from the early bronze age." Eleven hour lines radiate from the gnomon hole across two engraved day curves to the lower edge. The dial's top surface slopes downward about 15° from horizontal.

For $\varepsilon = 24°$, g = 217 mm,
$\phi_W = 46°$
$\phi_S = 45°$

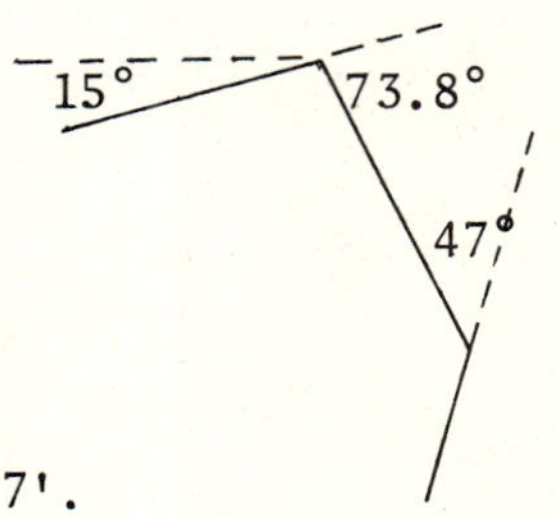

In making my calculations I have assumed that the summer solstice coincides with the lower edge of the cylindrical surface. Ionescu-Carligel assumes that the two engraved day curves are summer solstice and equinox. He obtains g = 166.7 mm, $\phi = 44°7'$.

Comanescu, S., "Dechiffrement et interpretation du cadran solaire de Cumpana (Dobrudza)," *Acta Antiqua Philippopolitana Studia Archeologica*, Sophia, 1963, pp. 181-84 (photograph).
Ionescu-Carligel [1], pp. 199-208 (photograph).
Ionescu-Carligel [2], pp. 119-37 (photograph).

Sundials with Multiple Faces

This section of the catalogue contains Greek and Roman sundials whose designs incorporate combinations of different shapes of shadow-receiving surfaces. For example, dial 7001G includes both spherical and planar faces. Rather than force a choice between earlier sections of the catalogue, I have elected to include dials like 7001G in the special section which follows. The dials in this special section are by no means the only examples with multiple faces. Dials like 1069G, 1072G, 1073, 1074G, 3049G, 3106G, 3107G, 3108G, 5001, 5002G, 5008G, 5020, and 5022G are composed of multiple faces of uniform shape. Assignment to a single earlier section of the catalogue involved no choice, and hence there was no need to segregate them.

The abbreviations used in this section may be found at the beginning of earlier sections: spherical, p. 122; roofed spherical, p. 194; conical, p. 219; horizontal, p. 323; vertical. pp. 339-41; cylindrical, p. 369.

7001G. Spherical Variant
Tenos, Greece
Archeological Museum, no. A139

Pentelic marble - Four separate shadow-receiving surfaces face the cardinal points of the compass on this dial. The base below the south-facing surface is flanked by two dolphins. The area below the north-facing surface is inscribed with a lengthy Greek dedication:

[Πάτ]ρα σε Κύρρος ᾿Ανδρόνικε δεύτερον
[῎Αρα]τον ἐν ζωοῖσιν ἄλλον ἔτρεφεν
[σὺ μὲ]ν γὰρ ἔγνως οὐρανοῖο παμφαῆ
[κύκλο]ν ταμέσ̣θ̣α̣ι̣ [σ]φαιρικάν τε πασσόφο[υ]
[τέχναν] ᾿Αράτου, [σὺ]ν δὲ καὶ λιπαυγέα
[ἔκλε]ι̣ψιν ἄστρω[ν] ἀτρεκῆ προθεσπίσα[ι]
[καὶ πᾶ]ν ἔπα[θλ]ο̣ν εἶλ̣[ες] αἰθεροδρόμων
[γραφ]ᾶι κελε̣[ύ]θ̣ους ὡρέων διαπλάσας
[παν]τὸς δ᾿ ἐπ᾿ ἔργο[υ ζ]ᾶλος, ἔνθα πολλάκις
[νίκασα]s ἁ φέρ[ει δ]ὲ̣ [τί]μιον γέρας
[πάτρ]α μάκαιρα [Κ]ύ̣ρρος Ἑρμίου τέκος,
[Μ]α̣κηδόνων ἕδεθ[λο]ν, ἅ σ᾿ ἐγείνατο.

Alternate restitutions have been suggested by H. Diels as follows: line 2, Εὔδο]ξον; line 5, διελεῖν] ; line 7, [καὶ νῦ]ν; line 9, [γνω]τὸς; line 10, [αφῖξο]σὰ.

H = 689 mm
W = 580 mm
D_{top} = 569 mm
D_{base} = 492 mm

South Face

Seven hour lines (3,4,5,6,7,8,9) and five day curves are preserved on the unrestored portion of this face. The configuration of lines suggests that the restoration as a roofed spherical dial is correct. Although this face of the dial is badly worn, the hour lines do seem to have once extended from the winter solstice to the summer solstice. The seasonal indicators c_1 and c_2, extending only between the fifth and seventh hour lines are marked:

wc_1 = 120 mm
c_1e = 95.5 mm
ec_2 = 122.5 mm
c_2s = 61.5 mm
sb = 34 mm

(VI to VII)
w = 45 mm
(V to VI)
c_1 = 76.5 mm
c_2 = 133.5 mm
s = 145 mm
(IV to V)
e = 107.5 mm

7001G. continued

πλειάδων δύσις
χειμῶνος ἀρχή

πλείας ἐκφανής
[θέ]ρους ἀρχή

d = 500 mm

α = 65°
β = 34°

The equinox is marked
[ἰ]σημερί[α].
The area between c_2 and s is marked
κύων ἐκφανής.
Three fragments have been restored (probably incorrectly) to the top of the sphere.

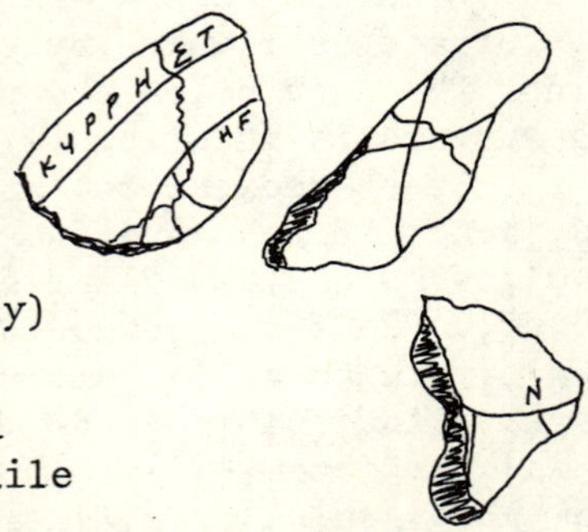

For ε = 24°, oe = 411 mm, and
r_w = 257 mm with ϕ_w = 37°, while
r_s = 220 mm with ϕ_s = 20°.
For r = 257 mm, c_1 corresponds to δ = -10°43'
and c_2 corresponds to δ = 13°44'.
At c_1 the solar longitude is Libra 28°; at c_2 the solar longitude is Taurus 6°. Geminus reports that Democritus observed the Pleiades on the fourth day of Scorpio; Euktemon observed the morning rising of the Pleiades on the thirteenth day of Taurus.

North Face

Eleven hour lines extend between the winter and summer solstices. Five other day curves, including the equinox, cross these hour lines. A groove and circular depression suggest the former existence of a gnomon projecting from the bottom of the spherical surface. The summer solstice is inscribed
τροπὴ θερινή.
The equinox is marked
ἰσημερία.

sc_1 = 10 mm
c_1c_2 = 29 mm
c_2e = 46.5 mm

(III to IV)
s = 50 mm
c_1 = 51.5 mm
c_2 = 53 mm
e = 57 mm
c_3 = 55 mm
c_4 = 60 mm

For r = oe = 218 mm, ε = 22°30'.
The inclined gnomon may have been placed in the plane of the equator.

7001G. continued

West Face

The winter solstice (marked τροπὴ χειμερονή, equinox (marked ἰσημερία), and a trace of summer solstice are visible. Six hour lines (as partially restored) extend from the winter solstice to the front edge. They divide the equinoctial line into the following segments measured from the base to the horizontal line near the top: 405 mm, 185 mm, 73 mm, 39 mm, 30 mm, 25.5 mm, and 17 mm.

we = 55 mm
e to front = 144 mm

$\alpha = 52°$

$h = (\text{XII to IX})_e = 72.5$ mm.
$\phi = 90° - \alpha = 38°$.

East Face

The winter solstice (marked τροπὴ χειμεθ[ινή]) and the equinox (marked ἰσημερί[α]) are visible. Five hour lines extend from the winter solstice to the front edge. The sixth hour line connects the winter solstice and the equinox. The hour lines divide the equinox into the following segments measured from the sixth hour to the horizontal line near the top: 185 mm, 70.5 mm, 39 mm, 34 mm, 24 mm, 16 mm.

we = 60 mm
e to front = 147.5 mm

$\alpha = 53°$

$h = (\text{0 to III})_e = 74$ mm.
$\phi = 90° - \alpha = 36.5°$.

Graindor, P. "Horloge d'Andronikos," *Musée Belge* 10(1906): 353-61 (photographs and drawings).
IG, XII, 5.2, 891 (drawings) and supplement p. 139.
Cf. Diels, pp. 172-73 and plate 13 (drawing).
Rehm, A. "Griechische Windrosen," *Sitzungsberichte der Königlich Bayerischen Akademie der Wissenschaften*, May 6, 1916, pp. 12-13, note 1.

7002G. Globe
Nafplion, Greece
Archeological Museum

Plates 62, 63, 64
Page 378

White marble - Three separate networks of lines have been engraved on this spherical stone. Two great circles on the sphere correspond to the meridian and horizon circles. A conical hole with base 100 mm in diameter pierces the globe at the lower pole.

radius of globe = 262 mm

Found by Carl W. Blegen at Prosymna in the Argolid.

Upper South Face

Plate 62
Page 378

Eleven hour lines extend between summer solstice (at top) and winter solstice (at bottom). Five day curves cross the hour lines. Thirteen shallow holes marked with Greek letters are distributed along the prime vertical circle of the globe:

ϛ Ε Δ Γ Β Α (ΙΒ) ΙΑ Ι Θ Η Ζ ϛ

The first and last holes are at the junctions of the horizon and the prime vertical. The middle hole is at the junction of the meridian and the prime vertical. The other holes intersect hour lines. The spaces between the day curves are indentified with Greek inscription as signs of the zodiac:

καρκίνος	δίδυμοι
λέων	ταῦρ(os)
παρθέν(os)	κριός
ζυγός	ἰχθύε(s)
σκόρπι(os)	ὑδρηχό(os)
τοξότης	ἀιγόκε(ρως)

tc_1 = 50 mm
c_1c_2 = 15.5 mm
c_2c_3 = 40.5 mm
c_3c_4 = 54.5 mm
c_4c_5 = 51 mm
c_5c_6 = 42.5 mm
c_6c_7 = 16 mm

(VI to O)
c_1 = 132,74,71.5, 67,68,52.5 mm
c_2 = 126,74,69, 69,71,53.5 mm
c_3 = 111,70,70, 70,75,56 mm
c_4 = 80,62,70, 69,79,65 mm
c_5 = 58,55,65, 59,77,98 mm
c_6 = 43,47,52, 55,62,141 mm
c_7 = 40,42,47, 52,58,156 mm

7002G. continued

Lower South Face

Plate 63
Page 378

Three great circles meet at the east and west poles. Eleven cross lines extend between the first and second great circles. Thirteen cross lines extend between the second and third great circles. Shallow holes mark the east and west points and the junction of horizon and meridian.

c_1c_2 = 79 mm
c_2c_3 = 80 mm

segments west to meridian:
c_1 = 238,75,44, 28,21,16 mm
c_2 = 111,88,76, 55,51,43 mm
c_3 = 88,69,71, 60,72,64 mm

North Face

Plate 64
Page 378

This configuration of lines closely resembles that on the concave surface of a roofed spherical variant dial. Eleven hour lines and three day curves are visible. A shallow hole marks the meeting point of the three day curves.

A two line Greek inscription is preserved below the north face:

῞Ηρης ἱροπόλος με θεῆς ἀνέθηκε Θαλεία
ἡλιακῶν ὡρῶν ἄγγελον ἡμερίοις

tw = 148 mm
we = 109 mm
es = 112 mm
s to horizon = 53 mm

(VI to 0)
w = 30,37.5,32.5, 30,30,28 mm
e = 55,63,58,65, 79,84 mm
s = 55,69,74, 80.5,110, 250 mm

Blegen, C. W., "Prosymna, Remains of Post Mycenaean Date," AJA 43(1939): 443F and figure 31 (photograph).

62

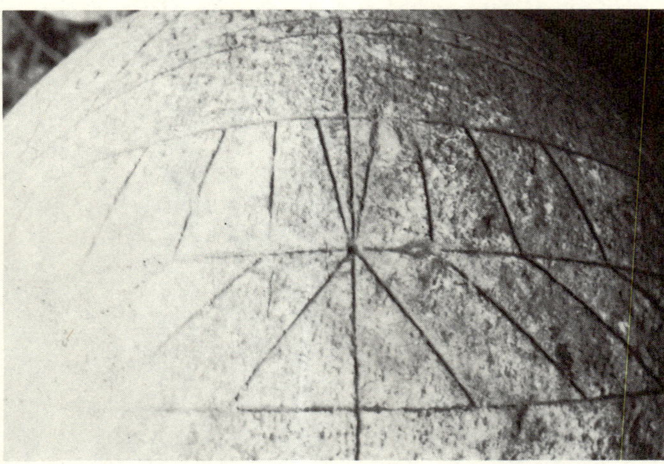

63

64

7003G. Conical and Spherical
Palermo, Italy
National Archeological Museum, no. 5587

Seven gnomon holes are visible in the top of the dial block. One is on the north-south axis near the north face. Two groups of three are positioned symetrically above the east and west faces.

H = 266 mm
W = 488 mm
D_{max} = 483 mm

White marble - Four shadow-receiving surfaces have been carved into this block: a south-facing conical surface, a north-facing conical surface, a west-facing quarter sphere, and an east-facing quarter sphere.

Although no acquisition information is available, the dial is marked Tindari 110, and the curator assumes that it was found at that site.

South Face, Conical

Seven hour lines are preserved (3,4,5,6,7,8,9). Three day curves are visible (winter solstice, equinox, and summer solstice). The area above the winter solstice is broken away. The hour lines do not continue beyond the summer solstice.

we = 54 mm
es = 116 mm
sb = 12 mm

(VI to VII)
w = 21 mm
e = 39 mm
s = 72 mm

For $\varepsilon = 24°$, $\omega = 39°$, $\phi_W = 42°$, and $\phi_S = 31°$.

$\gamma = 34°$
$\beta = 45°$

7003G. continued

North Face, Conical

Six day curves cross eleven hour lines. The hour lines do not extend beyond the upper and lower day curves.

For se = 43 mm, ew = 136.5 mm, and $\varepsilon = 24°$: $\omega = 49°$, $\phi_w = 40°$, and $\phi_s = 32°$.

gc_1 = 79 mm
c_1c_2 = 15.5 mm
c_2e = 43 mm
es = 136.5 mm
ss' = 11.5 mm
s'b = 7 mm
(VI to VII)
c_1 = 21 mm
c_2 = 24 mm
e = 27 mm
s = 42 mm
s' = 42 mm
$\alpha = 133.5°$
$\gamma = 40°$

West Face, Spherical

Five hour lines extend between the winter and summer solstices and cross the equinox.

For $\varepsilon = 24°$, r_e = 145 mm
r_s = 138.5 mm with $\phi_s = 28°$
while r_w = 148 mm with $\phi_w = 38°$.

tw = 68 mm
we = 58 mm
es = 62 mm

(VI to VII)
w = 27 mm
e = 38 mm
s = 40 mm

East Face, Spherical

Five hour lines extend between the winter and summer solstices and cross the equinox.

tw = 68 mm
we = 58 mm
es = 62 mm

(V to VI)
w = 27 mm
e = 38 mm
s = 40 mm

Cf. Drecker, p. 57.
Gatty, p. 41.

7004G. Cylindrical and Spherical — Plate 65, Page 382
Rome, Italy
Museo Nazionale Romano, courtyard

Marble - The dial faces are cylindrical and hemispherical. Two winged female figures hold the hemisphere in the lower position. A torch and scroll decorate each side of the upper cylinder. A stepped molding outlines the decorated areas and the shadow-receiving surfaces.

H = 318 mm
W = 351 mm
D_{max} = 86 mm

Found at the Villa Scipio near Herculaneum in 1769.

Top, Cylinder

Restoration has obliterated any trace of a gnomon hole.

H = 195 mm
d = 137 mm

Six preserved hour lines (morning hours) and the meridian cross three day curves.

tw = 24 mm
we = 24 mm
es = 94 mm

For $\varepsilon = 24°$, $\phi = 37°$ and h = 31 mm. The radius of a trace of the top edge is 70 mm.

Bottom, Hemisphere

Of the eleven hour lines crossing the three day curves, only the meridian extends beyond the summer solstice. The day curves are not parallel to the front edge.

d = 127 mm

we = 30 mm
es = 25 mm
sb = 23.5 mm

(VI to VII)
w = 11 mm
e = 18.5 mm
s = 24 mm

A remarkably similar dial with multiple faces, including a cylindrical surface, is now preserved in a private collection in Belluno, Italy (see 7005G).

Cf. Drecker, p. 41, figure 67 (photograph).
Gatty, p. 42 (drawing).

65

7005G. Multiple Faces
Belluno, Italy
Private Collection

Plates 66, 67, 68
Page 387

H = 320 mm
W = 393 mm
D = 210 mm

Luna marble - Eight hour line and day curve grids are engraved on five differently shaped surfaces. Two winged cherubs flank the cylinder at the top front of the dial. A crescent moon and three stars decorate the lower left corner in front; an eagle decorates the lower right corner. On the right side, the legs of a cherub running toward the back of the dial decorate the top; a piece of drapery in relief blows from back to front in the bottom section. An engraved line following the circumference of the circular section of the surface at the lower left front is the only decoration on the left side. A stepped molding outlines the various shadow-receiving surfaces and decorated areas.

Provenance unknown.

The arrangement of hour lines and day curves on all shadow-receiving surfaces of this dial indicate that its designer intended the front edge to lie parallel to the meridian line. A very similar combination dial (H = 318 mm), now preserved in the Museo Nazionale Romano, was designed to function with the front edge perpendicular to the meridian line (see 7004G).

Top Center, Cylinder

Plate 66
Page 387

H = 135 mm
d = 141 mm

te = 45 mm
es = 86 mm

Three gnomon holes are visible above the top left side of the cylinder. The larger hole has diameter 7 mm; the two smaller holes have diameters of 5.5 mm.

Five hour lines cross two preserved day curves. The measured distance from meridian to equinox along the top edge is 66 mm. The measured

7005G. continued

distance from the summer solstice to the equinox along the top edge is 53 mm.

For tw = we = 22.5 mm and $\varepsilon = 24°$,
h = 29.3 mm and $\phi = 37°16'$.
For a distance from meridian to equinox along top edge of 66 mm and r = 70.5 mm, h = 28.7 mm.

Top Right, Conical

A gnomon hole, 7 mm in diameter, is visible above the meridian line. Five preserved hour lines cross three day curves. The surface is one half of a regular south-facing conical surface.

There are no lines traced on the north-facing conical surface at top left.

H = 135 mm

we = 23.5 mm
es = 45 mm

(VIII to IX)
w = 7.5 mm
e = 12.5 mm
s = 18 mm

For $\varepsilon = 24°$, $\omega = 35°$.

Lower Left, Half-Truncated Cone

Lead remains in the 7 mm gnomon hole at the top right of the trapezoidal front section of the surface.

Six hour lines cross three day curves. The meridian is not engraved. The distance along the meridian line from the right boundary of the truncated cone to the summer solstice measures 32.5 mm.

W = 113.2 mm
$d_{smaller}$ = 64.5 mm
d_{larger} = 108 mm

se = 32 mm

(VI to VII)
s = 14.6 mm
e = 17.8 mm

7005G. continued

Lower Right, Half-Truncated Cone

Lead remains in a gnomon hole at the top left of the trapezoidal front section of the surface.

Six hour lines cross three day curves. The meridian is not engraved. The distance along the meridian from the right boundary of the truncated cone to the summer solstice measures 50 mm. The distance along the meridian from the winter solstice to the left boundary of the truncated cone measures 10 mm.

W = 113 mm
$d_{smaller}$ = 64.5 mm
d_{larger} = 108 mm

se = 35.5 mm
ew = 28 mm

(VI to VII)
s = 18 mm
e = 17 mm
w = 18.5 mm

Lower Center, Hemisphere

A gnomon hole, 11.5 mm in diameter, is visible below the lowest point of the front circular section.

Five hour lines cross three day curves. The meridian is not engraved, but day curves continue to the implied meridian at the front edge. Two parallel curves, c_1 and c_2 are engraved at an oblique angle in the right half of the hemisphere. The arc bs subtends an angle approximately equal to 15.5° at the center of the sphere; be, an angle of 37°; bw, an angle of 60.5°; bc_1, an angle of 40°; and bc_2, an angle of 62°.

d = 139 mm

bs = 22 mm
se = 22 mm
ew = 29.5 mm
bc_1 = 49 mm
c_1c_2 = 29.5 mm

(VI to VII)
s = 22 mm
e = 20 mm
w = 12.5 mm

7005G. continued

Back Right, Spherical — Plate 68

A gnomon hole 9 mm in diameter is visible at the lowest point of the spherical segment.

Five hour lines cross three day curves. The distance along the unengraved meridian between the lower left boundary of the segment and the summer solstice measures 89 mm.

d = 165 mm

se = 34 mm
ew = 32.5 mm

(0 to I)
w = 15.5 mm
e = 19.5 mm
s = 21 mm

r_m = 74.48 mm
For $\varepsilon = 24°$, r_w = 78 mm and
$\phi_w = 24°$ with $\phi_s = 32°$.

Back Left, Spherical

A gnomon hole is visible at the lowest point of the spherical surface.

Six hour lines cross three day curves. The meridian line is engraved. The distance along the meridian from the right boundary of the spherical segment and the winter solstice measures 29 mm.

d = 156 mm

se = 26.5 mm
ew = 36.5 mm

(0 to I)
w = 14.5 mm
e = 19.5 mm
s = 23 mm

For $\varepsilon = 24°$, $\phi_w = 45°$ and $\phi_s = 37°$.

Back Center, Spherical

The distance between cross lines on the right day curve is 34 mm. The distance between cross lines on the left day curve is 37 mm.

Portaluppi, Piero, *Gnomonica Atellana*, Milan, 1967, pp. 100-01 (photographs).

66

67

68

7006G. Cylindrical
Vatican City
Vatican Museum, Cortile Ottagono, no. 946

H = 495 mm
W = 450 mm
D = 320 mm

Fine-grained light gray marble - Parts of three shadow-receiving surfaces are preserved: left, right, and center. All engraved lines are wide and deep. The rough-finished back is curved to correspond to the front surface.

Left Front, Cylinder

es = 220 mm

The cross section of this cylindrical surface is non-circular. The preserved hour lines include 6,7,8,9,10,11. The ninth hour line extends beyond the winter solstice; all other hour lines are contained within both solstices. The winter and summer solstice do not extend beyond the eleventh hour line, but the equinox continues to the top of the dial. The summer solstice curve from the seventh to the eleventh hour lines consists of the following arcs: 102,56,40,35 mm. The equinox from the seventh hour line to the horizon consists of the following arcs: 40,30,30,30, 21 mm. The winter solstice curve from the seventh to the eleventh hour lines consists of the following arcs: 19,25,25,19 mm.

7006G. continued

Right Front, Cylinder

A gnomon hole is preserved in the top of the dial above the meridian.

we = 72 mm
se = 220 mm

The cross section of the cylindrical surface is noncircular The preserved hour lines include 1,2,3,4,5,6. All hour lines extend slightly beyond the summer solstice but terminate at the winter solstice. Neither solstice extends beyond the first hour line, but the equinox continues to the top of the dial. The summer solstice curve from the first to the sixth hour lines consists of the following arcs: 35,40,56,102, 129 mm. The equinox curve from the horizon to the sixth hour line consists of the following arcs: 21,30,30,30,40, 42 mm. The winter solstice curve from the first to the sixth hour lines consists of the following arcs: 19,25, 25,19,20 mm.

For $\varepsilon = 24°$, $\phi = 41°$ and h = 106 mm.

Center Front, Cylinder

Three hour lines (5,6,7) and three day curves are preserved. The cross section of the surface is noncircular.

we = 93 mm
es = 288 mm

(VI to VII)
w = 32 mm
e = 55 mm
s = 172 mm

For $\varepsilon = 24°$, $\phi = 41°$ and h = 136 mm.

Amelung, vol. 2, p. 106.
Cf. Drecker, p. 41.

7007. Multiple Faces
Present Location Unknown

The central shadow-receiving surface of this dial is a convex cylinder with eleven hour lines and three day curves. Ashby publishes a measured drawing by Andreas Coner which gives tw = 3 units, we = 5 1/8 units and es = 17 3/4 units. The central dial is flanked by two deviating vertical dials facing southeast and southwest. Each deviating dial has three day curves divided by five hour lines with a gnomon hole above the meridian. A calendar showing the longest day equal to fifteen hours ($\phi = 40°56'$) is engraved on the base below the dial faces.

Ashby, T., "Drawings Attributed to Andreas Coner," *Papers of the British School at Rome*, vol. 2, London, 1904, p. 33, no. 48 and plate 48 (drawing).
CIL, VI, 2306.
Cf. Michaelis, A., *Jahrbuch des Instituts*, 1891, p. 237, no. 187.

Sundial Fragments with Inscription

This section of the catalogue might more appropriately be titled "Some Inscriptions Relating to Sundials." For it is by no means certain that every piece included here represents a sundial fragment. Both entries 8004 and 8006 are suspect. The term ὡρολόγιον may, but need not, refer to a sundial. Since descriptive details of each dial are lacking, I have decided to emphasize the inscription here rather than include the shadow-receiving surface in an earlier section.

8001. Found on a piece of marble in a wall in Amastris, Turkey. The letters could date from the second century B.C. or later.

Ὁ πρὸς ὀρθὰς...
πρὸς τὴν ἐν...
ὄντα Θ̄ προ...
ἡ μεγίστη ἡμέ[ρα...
δυεῖν πεντεκα[ίδεκα?...
ἐτηρ[ή]θ[η?] ἐν τ...

Kalinka, E., "Aus Bithynien & Umgebung," *Jahreshefte des Oesterreichischen Archäologischen Institutes in Wein* 28(1933): beiblatt column 74, no. 24.

Wilhelm, A., "Inschriften zweier Sonnenuhren aus Amastris und Samothrake," *Jahreshefte des Oesterreichischen Archäologischen Institutes in Wein* 30(1937): 135-41.

***

8002. Found at Didyma. The inscription is part of a now lost sundial with spherical surface. The epigraphy dates from the early second century B.C.

[κα]ὶ πα[ρεδρεύσας ἐ]ν τῶι ἱερῶ[ι]
Ἀπόλλ[ωνι] Διδυμεῖ
Ἀπόλλων[ι Δι]δυμεῖ

Rehm, A., *Didyma, Die Inschriften*, Berlin, 1958, p. 115, no. 85.

***

Found built into the wall of the house of Hussein Tschausch on Kos. The block of stone contained a gnomon hole and an inscription dating from the econd or first century B.C.

Φί]λιππο Αὐτοφωντο[s] τὸ ὡρολόγιον Τύχαι
Ἀγαθᾶι καὶ Ἀγαθῶι
Λαίμονι καὶ τῷ δάμωι

ische *Forschungen und Funde*, Leipzig, 1899, 7.

8004. A small marble base ornamented with a relief margin above and below was found on Lemnos. It contained the following Greek inscription dating from the first century B.C.:

Δέξιππος Παγχάρου Παιανιεὺς
γενόμενος κοσμητὴς Θε[οῖς Μεγάλοις
τὸ ὡρολόγιον [ἀνέθηκεν]

Accame, S., "Iscrizioni del Cabirio di Lemno," *Annuario della Scoula Archeologica di Atene e delle Italiane in Oriente*, vol. 3-5, 1941-43, p. 99, no. 17, figure 18 (photograph).

***

8005. The following single line of inscription was found on a sundial from Oropus-Amphiareion, Greece:

[--τοῦ]κύκλου τοῦ ἐνγὺς τοῦ γνώμονος ἡ σκιὰ/
πορεύ[εται-]

Mitsos, T., "'Εργασίαι ἐν 'Αμφιαρείῳ τοῦ 'Ωρωποῦ," *Praktika*, 1951, p. 128.

***

8006. The following inscription was found on a white block of marble found near the temple of Asklepios at Pergamon (inventory number 1929.23):

Π Αἴλιος Ἰνγένους...
τὸ ὡρολόγιον σὺν τῇ [μηχα-
νοποιιᾳ κατασκευ[άσας ἐκ τῶν
ἰδίων ἀνέθηκεν

Wiegand, T., "Ausgrabungen in Pergamon (Asklepieion) 1928-32," *Abhandlungen der Preussischen Akademie der Wissenschaften*, 1932, no. 5, p. 33, no. 3.

8007. The following Latin inscription was found on a sundial near the temple of Hercules in Pompeii, May 11, 1765:

L SEPVNIVS L F SANDILIANVS
M HERENNIVS A F EPIDIANVS
DVO VIR I D SCOL ET HOROL
D S P F C

Mommsen, T., *Inscriptiones Regni Neapolitani Latinae*, Leipzig, 1852, no. 2227.

***

8008. The following inscription was seen by Cyriacus of Ancona on a sundial from Samothrace:

Τοῦ γνώμονος ἡ
[σκι]ὰ ἐπιοῦσα ἐπὶ τὰ-
s γραμμὰs ση-
μαίνει τὰs ὥρας
τοῦ ἐναυτοῦ καὶ
τῆs ἡμέρας.
Τροπῶν θερινῶν ἡ
πρώτη, ἰσημερι[ν]ῶν ἡ μέση,
[χει]μερι[ν]ῶν ἡ ἐσχάτη.

The back of this stone was apparently also inscribed:

(Στ)ῆσε Ὁδιός με (Αἴ)σωνος ἔχοντα μέρη/
δ(εκ)α(π)έντε,
ἥ(ε)λιον δὲ (γρ)άφων γνώμονα τόνδε δ[έκα].

IG, XII, 8, 240.

Wilhelm, A., " Inschriften zweier Sonnenuhren aus Amastris und Samothrake," *Jahreshefte des Oesterreichischen Archäologischen Institutes in Wein* 30(1937): 141-47.

Index of Sundials by Present Location

Index of Sundials by Provenance

Provenance Latitude	Catalogue	Plate
Samothrace, Greece 39°46'	1075 8008	
Sebaste (Sivas, Turkey) 39°44'	5023	
Side, Turkey 36°45'	3059G	
Siebengebirge, Germany 50°38'	3104G	
Tarracina (Terracina, Italy) 41°17'	1014	
Tenos, Greece 37°33'	7001G	
Tergeste (Trieste, Italy) 45°39'	2015	
Tibur (Tivoli, Italy) 41°58'	1012 4011	
Tomi (Constanta, Rumania) 44°10'	1065	
Traianopolis (Loutra, Greece) 37°51'	1004	
Tyndaris, Italy 38°08'	7003G	

The latitudes in this index have been taken from:
Johnston, W. and A. K., *The Unrivalled Classical Atlas*, Edinburgh, 1877?, and
The Times Atlas of the World, second edition, Boston, 1971.

General Index